THE 20TH CENTURY LEGACY

The Explosion of Technology
&
Humanity's Struggle with
Nature

by

FRANK J. WELCH

The 20th Century Legacy

The Explosion of Technology
&
Humanity's Struggle with Nature

Reflections of a Witness

to the Development of Science

And

the Explosion of Technology

COPYWRIGHT © Frank J. Welch
08/04/2016

All rights reserved. No part of this book may be reproduced in any form or by any electronic or mechanical means, including storage and retrieval systems, without written permission from the author except by a reviewer who may quote brief passages in review. Anyone who would like to obtain permission for any of the material in this book may contact:

Published by Frank J. Welch

University Village Thousand Oaks
3415 Campus Drive
Thousand Oaks, CA 91360

Printed in USA

-- ISBN 1543170382 --

Preface

"Action" seems to be a hallmark of the American culture; perhaps it is a characteristic of human nature. People are impatient and leap to conclusions before all of the evidence can be assembled. Americans assess an issue and take actions quickly to resolve the problem without undue study. When the problem is complex, however, often quick action addresses only symptoms and the basic issue remains unresolved. The rapid growth of technology has created a complex society that defies understanding. Impulsive action can be costly and dangerous with new technologies. The way of science, seeking truth and understanding, requires deliberative study.

Writers of the history of the Industrial Revolution have focused solely on the development of machines and their correlation with social and economic changes. There were other developments, however, that also effected the changes in society. These included the creation of energy and many new materials that made the machines possible.

It was also a time of nation building, colonial empires, and expanding global trade.

This history takes a new, expanded view of technology and its relation to science, government, economics, and ecology.

Introduction

Scientists say the universe was created sixteen billion years ago with a big bang that formed galaxies of stars of untold energy. Matter consisted mainly of hydrogen and helium gases. Some four billion years ago the earth was created in a "goldilocks" orbit around the sun, not too hot and not too cold. The earth was unusual in that it contained the element carbon that could combine with hydrogen, oxygen, and nitrogen to form many substances. Perhaps one billion years ago large hydrocarbon molecules formed that had the ability to reproduce themselves and living creatures formed. Then over many millions of years, plants, reptiles, fish, birds, and mammals arose. About one million years ago the first humanoid ancestors appeared who evolved into humans some fifty thousand years ago.

Over many centuries early humans learned to make use of their environments. They tamed fire and domesticated animals. Their mastery of their environments progressed through the Stone, Bronze, and Iron Ages. In the Renaissance of learning in Western Europe five hundred years ago modern Science stirred and struggled through infancy and adolescence while exploring the natural world. Then during the Industrial Revolution, that began two hundred fifty years ago, steam engines were invented to power factories, ships, and trains. Accompanying this development of power was the expansion and reach of communication processes. Society changed forever in the 20th century as science reached maturity and discovered the secrets of the natural world. Powerful machines and improved communication methods created a whole new way of living.

This is the story of the manufacture and utilization of energy to produce power, the creation of new materials, the mastery of science, and the effect of technology on society. It is a history which looks beyond the surface of events to underlying causes.

The story begins with the creation of energy in the waning years of the 18th century, probably, the most important discovery of the Industrial Revolution. It passes through the development of motor powered machines in the 19th century. Then came the mastery of science, the development of powerful engines, and the creation of remarkable new materials in the 20th century. It culminates with the creation of electronic devices leading to today's world of instant communication, robotic controls, virtual reality, and artificial intelligence.

We are living in a dynamic world of pixels and opinions masquerading as facts. To understand what is happening we need to step back from the details of life and consider a broad overview. One can't see the forest while wandering among the trees!

In addition to many benefits, technology has created problems that challenge the very existence of modern society. These dangers include the unthinkable threat of nuclear warfare, a growing world population devouring natural resources, the economic instability of governments, automation and specialization eliminating jobs, and the loss of personal security and freedom. Technology has created a complex, confusing, dangerous, and chaotic world that has upset long established cultural values and social structures. Preserving civilization from the destructive side effects of technology is the critical issue facing humans.

This "outside the box" approach gives a much different picture of the way technology impacts society than the usual view. It is more than a history; it is a critique of the workings of technology that exposes the fallacy of many popular beliefs. It explains why the 20th century was unique in human experience and why we should not expect to replicate the unusual economic growth in the future by new scientific discoveries.

This is the story of humanity's struggle with nature. In the 20th century, humanity discovered the secrets of the natural world but learned little new about human nature, the innate behavior of individuals. Humanity's conflict with human nature is ages old, but technology has magnified the consequences of the conflict by bringing the cultures of the world into intimate contact. This is a picture taken from the perspective of a scientist who witnessed the technical miracles of the 20th century.

Contents

Preface

Introduction

Chapter		Page

Part I The 20th Century

1	Background	1
2	Energy & Power	11
3	Materials	25

Part II Technology's Effect on Society

4	Technology & Economy	33
5	Technology & Government	45
6	Democracy - California Style	66
7	Technology & Security	74

Part III Ecology

8	Pollution	87
9	Conservation	100
10	Global Warming	105

Part IV Human Nature, Science, and Technology

11	Human Nature	119
12	Science & Technology	132

Part V The 20th Century Legacy

13	20th Century Legacy	157

References	178
Acknowledgement	179
About the Author	180

Part I

The 20th Century

The Creation of Power, Science,

and a Mighty Nation

1. Background

The year 2000 came quietly after fears of wide spread computer system failures proved unfounded. But soon disasters struck: stock market crash, terrorist attack in New York, real estate crash destroying wealth for many, rising unemployment, and stagnating economy. Today we are faced with wealth concentrated in the few, a suffering "middle class", under employment, increasing numbers of retired people, growing national debt, and fear that human activities are affecting Earth's climate. The political system is paralyzed over differences of opinions of how to resolve these problems.

Evolution of Technology

When immersed in today's complex technical society, it is hard to realize how extensively technology has changed the way people live and how quickly this has happened. Early civilizations considered earth, water, wind, and fire to be basic elements and combined them to create tools, utensils, shelter, and weapons. By heating selected minerals, they created mercury, tin, copper, lead, and iron. They made pottery and adobe bricks from wet clay and glass by heating sand. Skin, fur, and bones of animals were used for clothing, tools, and parchment writing material. Trees and plants were used for construction of shelters, wagons, and ships. Wind and water provided the only source of power beyond the strength of people and their animals. Communication was limited to face to face contact and messages delivered by couriers.

Power

Then early in the 18th century, after thousands of years of human civilization with little structural change, the steam engine was invented and society began to change. Firewood was becoming scarce in England for heating; and coal began to replace it. Coal was much easier to handle and more efficient than wood but mines often flooded. The steam engine was invented to pump water from flooded coal mines. This early machine was massive and not very efficient. It was not until the late 1700s that a high pressure steam engine was created that was much more efficient. High pressure steam engines were developed to power ships, locomotives, and factories in Europe and in the U.S. By the middle of the century, steam ships were replacing sailing ships, and railroads were replacing canal and river boats. Toward the end of the century, oil replaced coal as fuel; thereby, extending the range of steam ships and trains. More efficient steam turbine engines were developed that are still used.

Background

In the late 1890s engines were developed that used the gases produced by combustion directly for power. These internal combustion engines were smaller, more compact, and more powerful than steam engines. Such engines led to the development of automobiles and airplanes in the 20th century, revolutionizing transportation and the way we live. In the waning years of the 19th century, electric motors were developed for use in factories and appliances. Nuclear energy was harnessed in the 1950s to generate electricity and to power ships.

In the 18th century people accidentally discovered how to make energy; in the 19th century they discovered how to make engines to use energy; and in the 20th century they learned what energy is and built big engines to land men on the moon. Power changed everything as cars and airplanes soon shuttled people around the world.

Communications

As generations of inventors and engineers developed machines to generate power in the 19th century, other inventors were developing technologies for enhancing communication. Telegraphy, photography, and later radio were European inventions that were exploited in the U.S. as the country expanded across the vast North American continent. In the first half of the 20th century, telephone, radio, and television systems were developed to connect the nation. Soon communications spread across the seas to Europe. As the 20th century elapsed, computers and the internet spread the cyber world to individuals throughout the world.

Science

Science stirred as the 20th century began and bloomed during the latter half of the century. In the 1950s it was noted that 90% of the scientists who ever lived were then working. These modern scientists explored the far reaches of the galaxies and characterized the structure of atoms. Use of the entire electromagnetic spectrum was mastered, and much was learned of the nature of life itself. Many new materials, created from coal and oil, made modern life possible. New ceramics, metals, alloys, and composites became the components of modern machines and electronic devices. Development of sulfa drugs in the 1930s ushered in the pharmaceutical industry that today provides the medications for cure and prevention of diseases. Radiation from electron beams, lasers, X-rays, microwaves, and the sun itself is now finding wide spread commercial utility. The structures of the basic building blocks of life, such as stem cells and genes, are being manipulated to develop better medicines and food crops.

Technology has nearly eliminated plagues, famines, and infant mortality. Society now expects the generation of new technologies and new products to continue as the new norm to spur economic growth and prosperity.

Background

Creation of Nations

Accompanying the remarkable technical developments were great social and political innovations. The changes began mainly in Western Europe in the 19th century and spread throughout the world. Autocracies, kingdoms, and dictatorships were replaced by republics and democracies. Colonial empires such as those of Spain, England, Germany, France, and Turkey disappeared, to be replaced by new, independent nations. Arabia, a romantic land of sheikdoms and the fable of Ali Baba, was divided into the kingdoms of Saudi Arabia, Iraq, Syria, and several emirates. Capitalism and Communism evolved in economic conflict.

As new nations matured and prosperity grew, tribal influences waned. Forests and their coveted wood species were destroyed; habitats for wild animals and mineral resources were depleted. The world population is now growing to enormous numbers with much of the growth taking place in growing mega cities. Islam has become a dominant influence in western civilization as Christianity's hold weakens. Several Muslim sects now are in violent conflict to impose their views on large masses of people in the Middle East and Africa, even threatening to spread their values to Europe and the U.S.

Before the Industrial Revolution, the measure of strength was manpower, and war was personal, hand to hand. Now machines are the measure of strength. The disruptive effect of these dynamics on the way we live is probably the greatest social change since humans learned to utilize fire and invented the wheel.

Life in The United States

Steam powered ships and trains speeded travel and movement of commerce in the 19th century, and steam engines fostered manufacturing. In his book "The American Railway" in 1889, Thomas Curtis Clarke observed that "the world of today (1889) differs from that of Napoleon more than his world differed from that of Julius Caesar." Steam powered machines, however, made little change in the way people lived on farms, in villages, and in cities.

Most of the social changes we enjoy today in the U.S. occurred in the 20th century and continue to evolve. Automobiles and airplanes made possible by the development of internal combustion engines at the turn of the century provided unprecedented mobility for individuals and families. Automobiles and networks of highways led to the westward shift of population and our present suburban style of living. To paraphrase Thomas Clark, the world today differs more from his world than his world differed from that of Julius Caesar.

Background

Many nationalities, races, and cultures intermixed as never before. The wide-spread practices of slavery and indentured workers faded as women's suffrage, equal opportunity, individual rights, and labor laws were adopted. Racial heritage and diversity are celebrated now in the U. S. while integration of diverse cultures remains elusive. Deviant life styles and atheism are widely advocated and gaining political clout throughout the nation. Marriage is no longer limited to union between the sexes.

Prior to World War II, jobs for women were limited to teaching, stenography, clerking, and nursing. The war provided openings for women in many factory jobs to replace men who were called into the military. "Rosie the Riveter" became the symbol for job openings formerly restricted to men. Now, equal employment opportunity is the law of the land.

Homes got much bigger with added closets and larger wardrobes, indoor plumbing, central heating, and air conditioning. Bath tubs, used at least weekly, were replaced by more frequently used showers. Household soaps changed in form from bars, to powders, and now mainly liquids.

Family size decreased, and family structure changed with multiple wage earners and less in-home family dining. People now enjoy more free time; life styles are more sedentary. In the past children often followed the careers of their parents; father & son enterprises were common. Now children go off to college, acquire a profession, and follow independent career paths.

Clothing styles have changed with the availability of new fabrics. Standard wear for men no longer includes hats and three piece suits made of wool and cotton. Ties are worn only for special occasions. Standard wear for women now include pants and skirts instead of ankle length dresses and petticoats. Clothing for both sexes are made of many synthetic and natural fibers and fiber mixtures. Active wear for both sexes consists of shorts and T-shirts.

In the past people were concerned primarily with making a living. Medicines were mostly home remedies derived from plants and minerals that had been used for generations. Doctors made house calls. Since the 1950s there is a growing emphasis on health, fitness, and longevity. There is concern for food composition and quality, diets, and fitness. Medical emphasis seems to have shifted focus from cures of diseases and injuries, to prevention of illness and to cosmetic treatments for better appearance. Much of the increasing medical costs are spent to extend life of the aged. There is increased emphasis on bodily safety. Many new safety laws have been enacted to protect workers and the public from harm.

Less labor is needed to provide basic needs. Opportunities for unskilled labor are diminishing as new industries employing electronics have evolved requiring special worker skills and training. Machines that once required repair and maintenance have been replaced by lower cost versions that need little maintenance and often are scrapped rather than repaired. These factors together with increased life expectancy have made under-employment pervasive. Our economy is now powered by credit and closely tied to that of other nations. High debt has become an accepted fact for individuals as well as governments.

Before technology's impact, education emphasized mastery of arithmetic, penmanship, and reading. People were well-versed in the classics, mythology, and Christian doctrine. Latin was the language elective in schools. Before television and the internet, people listened to radio and read books, newspapers, and magazines. In the 1800s college curricula stressed agriculture, engineering, teacher training, military training, and religion. Liberal arts, business, law, medicine, engineering, teaching, and sports seem to be the main focus of colleges in the 20th century. As society has changed, the college focus has evolved to stress communication, business, arts, and science. Education standards for employment evolved from literacy in the 19th century, to high school in the 20th century, and to college now.

Technology has created a society of specialization in all fields of endeavor that makes understanding of common issues more difficult and reduces employment opportunities for the unskilled. Values have changed -- it appears our constitutional right to individual freedom of opportunity has evolved into a right to share in the national prosperity. Personal freedom is curtailed in favor of conformance to popular practices. The earth shaking conflicts of the 20th century and before are now scrutinized critically from the perspective of current values.

The increasing complexity of society and growth of government have created a growing need for lawyers and a litigious U.S. society. Legal defense by corporations and professionals adds to medical and business costs. Some lawyers have extended their practice from defending clients' rights to extracting wealth from the more prosperous; class action law suits were created for this purpose. Cost of justice for felons has risen as courts and jails try to insure that the guilty are not mistreated. Many new federal and state laws are enacted every year to make additional misdeeds crimes that are punishable by fines and incarceration. The U.S. leads the world in numbers of prison inmates. Perhaps overcrowded jails are more the result of too many laws forcing compliance of behavior than increased felonious attacks on citizens.

2. Energy & Power

Early humans learned to use fire for heat and for light. They found many uses for the materials produced by combustion. Perhaps, use of fire marked the beginning of civilization. It was not until the 18th century, however, that humans learned to create power from fire – the invention of the steam engine. Steam from boiling water was harnessed to propel pistons or turn screws. Historians consider the invention of the steam engine to be the beginning of the Industrial Revolution. Economists consider technological innovations like the steam engine to be the driving force of economic growth. The underlying importance, however, is that the steam engine was the first time power was created from the energy released by chemical reaction.

Of course, that could not have been understood until chemistry evolved as a science some two hundred years later. The significance of the discovery that energy released in chemical reactions could be harnessed to produce power, has not been recognized by historians nor economists.

We seldom think much about energy, it seems to be there when we need it and we are distressed when it fails. Energy is used in many ways: mostly, heating, lighting, and powering machines. Our bodies produce energy to lift a cup of coffee and to rise from a chair. Energy is the foundation of life and makes possible the technological innovations of modern society. We all know what energy does, but few recognize the significance between energy creation and energy transmission. There are many ways to transmit and deliver energy, but energy is created by just three processes: the impact of moving masses, molecular chemical reactions, and nuclear reactions.

Generation of Energy

Kinetic Energy
The force generated by moving masses like wind and flowing water is kinetic energy. It has been used for centuries to run mills and fill sails. It is also the force generated by the impact of a missile, a hammer, or a fist. Kinetic energy was the only source of power available before the Industrial Revolution beyond the strength of humans and their animals. Strictly speaking, kinetic energy is a delivery process - energy is needed to propel the mass.

Chemical Energy
Energy is created to various degrees in all chemical reactions. Humans first used chemical energy when they made fire by burning twigs. They welcomed the heat, but

had no way to harness it to power mechanical devices. Centuries ago the Chinese invented "black powder" (a mixture of charcoal, sulfur, and salt peter), which was used over the years to launch projectiles from guns and cannons. But the energy was released in a blast preventing its use to power machines.

For the last hundred years, fossil fuels have been the primary source of energy for the industrial world, generating electricity, heating homes, and fueling motor vehicles. Hydrocarbons, the energy component of fossil fuels, became the preferred fuel of the 20th century because the energy released is very high and the rate of energy release is controllable. Hydrocarbons are plentiful and more efficient than fuels having other chemical bonds such as alcohols and biofuels. Hydrocarbons are available as liquids and gases which are convenient for engine operation. The discovery and utilization of fossil fuels are the most important, but unrecognized, innovations of the Industrial Revolution. Modern society cannot exist without hydrocarbons.

Nuclear Energy

Nuclear energy was created in the heat of the conflicts of World War II. In one of the greatest scientific achievements of all time, a bomb was created of exceptional power. Massive amounts of energy are created by nuclear fission which converts mass to energy. This discovery was an exception to the widely held scientific principle that matter cannot be created nor destroyed. After the war, nuclear power was harnessed to produce electricity and to propel submarines and naval ships.

Electrical Energy Production

Energy may be transmitted through the air by radiation of heat and light from power generating sources such as fires and the sun; and it may be conducted thermally and electrically through materials. Buildings and homes have long been heated by conduction of steam and hot water through pipes and by radiation from stoves and fires. Generally, energy cannot be conducted very far. A major discovery of the 19th century was that electricity could be conducted through wires for considerable distances to provide light and to power electric motors. It is important to realize that electricity is a conductor of energy and not an energy source.

Electricity has two technical limitations: power is lost when transported long distances over wires and it can be stored and retrieved only in limited amounts. Electricity is most practicable when used near the generation source. Coal, gas, and nuclear electric generators can be built near where electricity is needed; but large wind and solar electric generating processes must be built where the energy is and electricity transported, often long distances to where it is needed.

Energy & Power

Fossil Fuels
Coal and gas are the lowest cost and the primary fuels used throughout the world for generating electricity. Due to large amounts of impurities (coal is combustible rock containing greater the 50% carbon) coal plants create considerable waste and air pollution from sulfur dioxide, nitrogen, and ash (acid rain) that require extensive pollution abatement processes. Gas, a much cleaner burning fuel (it is mostly carbon and hydrogen), sets a high target for versatility and economic efficiency for other generation methods to achieve.

Nuclear Energy
In the years following WWII nuclear fission of uranium was thought by many to be the panacea for power generation and many electric power plants were built in the U.S. and around the world. Large naval ships and submarines were fitted with nuclear engines that allowed them to go to sea for months without need for refueling. Nuclear fission remains a cost effective process for generating power. However, because of the weight of shielding necessary to contain damaging radioactive radiation, nuclear reactors are not well-suited for land or air transportation uses.

Public awareness of the great destructive power of nuclear explosions and fear of uncontrolled runaway nuclear reactions have cooled public support for construction of new nuclear power plants. While there have been several nuclear accidents in the forty years of nuclear power generation, most damage has been contained to the reactors and nearby areas. Probably, there has been more damage and injuries from mining coal and shipping train loads of petroleum liquids around the country than from nuclear power plant operations. However, handling and disposal of spent reactor fuel, which remains highly radioactive for centuries, is a major issue. A nuclear fusion process that creates little radioactive waste could replace the current nuclear fission process if research succeeds. Maybe this will be the answer in the next century.

Renewable Sources

Spurred by environmentalists, governments have decided that electricity should be produced "naturally" by solar radiation, wind, or water even though for the foreseeable future these technologies are more costly and less suitable for present needs than production from coal, gas, or nuclear fission. Governments are requiring that electric companies use an increasing percentage of power from renewable processes, thereby increasing electric rates for all. Tax benefits are provided to investors in wind and solar ventures, to companies making solar receptors, and to buyers of solar systems. Government subsidies make it difficult to access the true competitive costs of renewal processes.

Water

Electricity was first produced in quantity from waterfalls and the process has a long history of economic use. Production is continuous as needed, day and night. Suitable sites for dams on rivers are limited, however, and often remote from customers. Further, environmentalists are pressing for removal of dams to return rivers to their previous natural condition thereby reducing further the availability of sites for power generation. There is now considerable interest in harnessing the energy from ocean tides. If tidal variability and logistic issues can be resolved, tidal power could become important.

Wind

The output from wind turbines is dependent on the velocity of the wind and the size of the sails on the turbines; so production is variable. Power production locations are restricted to geographical areas where wind blows consistently. Operating costs are said to be relatively small. Electricity is produced now from wind turbine "farms" in the U.S. and Europe although total output is still minor.

Solar

In the 1970s solar panels collected the heat radiated from the sun to warm water. High installation and maintenance costs for the receptor panels and operation limited to warm, sunny days, led to the demise of the program at substantial cost to governments and to pioneering customers including this author. The program was revived in the 1990s after silicon chips were developed to convert solar radiation to electricity directly. High costs of manufacture of solar panels and low efficiencies have slowed implementation, but improvements in performance of silicon chips make the process attractive. Since operation is limited to daylight hours, users have to retain service from local utility companies for use after dark. Large battery installations are now under development to store energy for use after dark, thereby freeing the process from support by electric utilities. Large batteries create a new economic and safety issue, however.

Solar energy is spread thinly across the Earth surface; so it is necessary to have large panel arrays facing the sun for large production. Installation sites must be carefully selected to provide maximum sun exposure; so large power plants require a great deal of space away from the urban centers where the power is needed. Consumer and production experience is still quite limited so economics remain uncertain. Solar is best suited to special applications and not to wholesale production of electricity replacing fossil fuels and nuclear.

Geothermal

Heat from the interior of Earth rises near the surface in many places and has been used for many years in limited amounts for heating. Currently, geothermal heating is being used in 70 countries around the world, and geothermal electricity generation is practiced in 24 countries.

The U.S. is the largest producer of geothermal electricity, mostly from California's Geysers field of wells and power stations. Geothermal energy is produced from deep wells that tap superheated water to feed electrical generators on the surface. Only a small percent of U.S. and world electricity comes from geothermal processes. Like the other renewable energy processes, operating costs are relatively low; but capital costs for exploration, drilling, and power generation are high.

Electric Vehicles
Electricity is the process of choice for urban trains, trams, and buses which can be connected to electric wires and rails. The lack of exhaust gases is a great asset for travel in densely populated areas. Mobile vehicles that cannot be attached to wires, however, must rely on electric storage batteries which severely limit performance.

Suburban society is based on motor vehicles powered by internal combustion engines that have been optimized for use of gasoline; no other source of power is equally effective. Electric motors can provide performance competitive to internal combustion engines, but they must drag along heavy batteries that reduce pay loads and require frequent recharging. Battery power is optimal for small vehicles with short range, such as golf carts, and factory vehicles. Electric vehicles designed for short distance, urban use seem attractive where exhaust fumes are objectionable, such as cabs and delivery vehicles. Great progress has been made in battery performance over recent years, but until technology is developed for storing large quantities of power electrically as well as gasoline does chemically, electric vehicles will be competitive only for special uses.

A process under development for overcoming the storage limitations of batteries for electric vehicles uses a Hydrogen Fuel Cell. The process combines hydrogen and oxygen from the air in a unique electrical cell to produce electricity. Since the only byproduct is water, it is environmentally friendly. The process is being tested in fleets of buses and automobiles; large scale trials in automobiles are under way in California and hydrogen filling stations have been installed in several parts of the state to evaluate handling hydrogen and performance in vehicles. Major limitations of fuel cells are comparatively low efficiency and high cost. The hazards of storage and handling hydrogen, which is highly flammable, particularly in large amounts, are also a concern. Hydrogen can be made by electrolysis of water, but the economics are unattractive compared to the usual production from petroleum. So hydrogen fuel cells do little to replace fossil fuels.

Another version of fuel cells that avoids handling hydrogen is an "electrolyte fuel cell" which employs two tanks of electrolytes that are pumped through a cell with a central membrane. Although test cars have been built using this process and hyped at auto shows, performance is not competitive with present lithium batteries. Electrolyte fuel cells are still at the early research stage.

<u>Hybrids</u> – Possibly the most practical present process for family cars is use of electric motors for power combined with small internal combustion engines to recharge the batteries for extended cruising range.

Petroleum Reserves

In the midst of the Cold War, great concern developed over the reliable sourcing of oil to supply the rapidly growing U.S. economy. Demand for oil was growing faster than U.S. production, so that considerable oil had to be imported. Then the oil-rich nations of the Middle East formed a cartel, OPEC, to control world oil markets. With uncertainties in the markets the Department of Energy was created in 1977 to attain energy independence and to establish reliable petroleum reserves. The petroleum industry continued to explore for new sources of gas and oil on land and sea and to improve recovery from existing fields. Subsequently, oil reserves expanded greatly around the world.

At the time that the anti-pollution message of the environmentalists was gaining political strength, oil spills contaminating sea shores, and fiery spills from tanker trucks and trains caused substantial ecological damage that was slow to heal. The nation began to resist expansion of drilling in coastal waters and in the Arctic. The natural reaction of people is to discard things that cause problems instead of managing the problems. So as world tensions eased and ecologists pressed to curtail use of fossil fuels, the focus of the Department of Energy was changed to replacement of fossil fuels by development of renewal energy sources.

Biomass
Government encouraged the development of fuels from renewable plant sources to replace petroleum. Al Gore considered energy generated from biomass to be one of the best ways to reduce carbon dioxide emissions.[1] Although biofuels could contribute to energy independence, they do nothing to reduce emission of greenhouse gasses. Biofuels are inherently less efficient than fossil fuels but still produce as much carbon dioxide. Never the less, Government has required that gasoline contain ten percent alcohol made from corn even though alcohol degrades engine performance and increases costs. A tax was levied on imported alcohol and a subsidy was paid to distillers making alcohol from corn. These subsidies led to construction of new factories to process corn for fuel with consequent higher costs for corn food products and animal feed.

Using agricultural lands for energy creation is an expensive and inadequate way to produce the large quantities of fuel needed for motor vehicles. Enormous acreage of arable land would be needed to produce enough fuel to supply U.S. consumption. With a growing world population, we will need all of the arable land and water available for food and clothing fibers. The high economic costs and futility of requiring use of biofuels was summarized by Thomas Donlan in a Barron's Editorial.[2] Al Gore later acknowledged that using alcohol from corn for fuel was a mistake.[3]

Sea plants, that would avoid the arable land criticism, are being investigated for biofuel production, but investigations are still in the exploratory stage.

Remarks

Energy is the foundation of civilization. Too little and life reverts several centuries; too much and civilization is destroyed. Exploration and enhanced recovery processes have greatly expanded known reserves of fossil fuels around the world; the U.S. is particularly rich in hydrocarbon reserves. Never the less, deposits of fossil fuels are finite, so effective management and conservation of supplies is critical for future generations.

For many years governments have been pushing development of energy processes that emit no carbon dioxide to replace fossil fuels. Financial incentives to manufacturers and to customers together with regulations that label exhaust components toxic to the environment have been imposed to encourage adoption of clean energy technologies. As yet, customer acceptance has been limited and profitable commercial success lacking. Electricity produced by wind and solar processes is promising for selected end uses, but economics remain uncertain and production volumes are limited by space considerations. Electricity is much costlier for heating buildings than natural gas and so is seldom used. Improved efficiency of electricity generation should expand uses for heating. Petroleum fuels will remain optimal for motor vehicles and air planes until a process for making energy is developed that is superior to chemical combustion.

There are several processes for generating energy and each should be used where it is most advantageous. However, forcing adoption of electricity to replace fossil fuels in motor vehicles before the technology is fully developed and acceptable socially and economically, imparts a very great cost on society and complicates solutions to protection of the environment and the best use of natural resources.

Government direction to adopt clean energy is ill advised - a costly misdirection of technical and commercial resources. Clean energy processes are adequately defined to remove subsidies and allow commercial enterprises to develop profitable applications. Effective use of the various kinds of energy is a major challenge in the years ahead.

3. Materials

Exploration of the unknown world was the hallmark of the 16th and 17th centuries, followed in the 18th century by the exploration of the new lands and their flora and fauna, and by a growing international trade. Research to determine the composition of materials proliferated in the 19th century, highlighted in 1869 when Mendeleev published the Periodic Table of the Elements. By then most of the ninety-two stable elements had been isolated and characterized. Most studies by that time had dealt with inorganic materials such as, minerals, metals, and ceramics. Emphasis then shifted to characterization of the more complex organic materials derived from carbon, such as wood, plants, and coal. The science of chemistry was taking form.

Cellulose was isolated from wood chips and from cotton in the middle of the 19th century and made into Celluloid plastic, and Rayon textile fiber. Cellophane wrapping film followed in the years before World War I.

Materials from Coal

As use of coal for heating and fuel for steam engines grew in the early 1800s, coal was also processed to make coke (carbon) to replace charcoal in the production of iron and steel. Coal was processed further to produce coal oil for lamps, and illuminating gas for lighting city streets. In the closing years of the century, graphite electrodes were created for electric arc furnaces that produced temperatures exceeding 2,000 degrees F. for smelting metals. This was an important use for newly available electricity. Graphite electrodes also were used for dry cell batteries. Calcium carbide, an important intermediate for making acetylene, was produced by heating coke with lime in an electric arc furnace. Acetylene gas was produced by adding water to calcium carbide. Lamps burning acetylene provided light for miners and for early motor vehicles. By the turn of the century, acetylene was used to make many other chemicals.

The coking-process produced useful byproducts including such gases as hydrogen, ammonia, and hydrogen sulfide together with residual coal tar. The coal tar industry arose with the separation of these "aromatic" compounds that served as the basis for making other chemicals. Many products were developed that remain in use today, including solvents, antiseptics, dyes, and Bakelite - the first industrial plastic. Germany became the center of the coal tar industry and the chemical leader in Europe in the waning years of the 1800s. The first coal tar plant in the U.S. was built in 1916.

Materials from Petroleum

The petrochemical industry arose from a project established in 1914 at Mellon Institute in Pittsburg to find a way to make acetylene from natural gas. The sponsor was Presto-o-Lite Co, a manufacturer of calcium carbide and a division of the newly created Union Carbide and Carbon Corp. George Curme, a young chemist returning from post-doctoral studies in Germany, was assigned the project. He failed to make acetylene, but he did develop and patent a process for producing ethylene in 1919. Four years later the Carbide & Carbon Chemicals Company was established in South Charleston, W.VA., to produce ethylene and several ethylene derivatives including ethanol and glycol. The South Charleston plant continued in production for over 50 years. Union Carbide built a plant on the gulf coast of Texas in 1941 to produce the same basic chemicals from naphtha, a liquid petroleum fraction. Ethylene, propylene, and butadiene, soon became the building blocks for the massive petrochemical industry.

Many new materials were created in the years leading up to World War II as understanding of chemical composition and processes evolved. Several early plastics, adhesives, coatings, medicines, and anesthetics were introduced. The War provided a stimulus to develop substitutes for scarce raw materials, such as natural rubber, as well as antimalarial and antiseptic drugs. The industry grew rapidly in the 1950s to 1970s with invention of the many materials that are essential elements for modern life. Pharmaceuticals, many kinds of plastics, textile fibers, films, protective coatings, elastomers, and adhesives, all tailored for special applications were developed.

Now chemicals touch nearly everything in modern life. The American Chemical Council reported that chemical products find their way into 96% of all manufactured goods. Petrochemicals are produced in more than 100 factories in 50 countries. Annual U.S. chemical industry production is about $760 billion, one of the largest industries. The bulk of chemical production goes into plastics for molding devices, wrapping films, textile fibers, protective coatings, and adhesives. Often plastics replace wood, metal, paper, natural fibers, rubber, and glass reducing costs and improving performance. In other cases, plastics provide unique performance features that expand capabilities.

Other 20th Century Materials

Many other materials that are essential components of modern machines and devices were developed since 1900. Examples include aluminum, titanium, iron alloys, and composites of resins with glass or carbon fibers. Silicon semiconductors and optical glass fibers and sheets are essential components of electronic systems. Compounds of metals, such as platinum, lithium, radium, uranium, and rare earth elements are essential components of electronic devices.

Age of Carbon

Modern society is founded on the utilization of carbon. Unlike any other element, it combines with itself and with other elements to form more compounds than all of the other elements together. It is the foundation for all living things. The utilization of carbon in the 20th century is the most significant mastery of new materials for civilization since the Iron Age.

Although coal and petroleum are both composed largely of hydrocarbons, their chemical structures differ as do the products made from them. Both kinds of products are important components of the pharmaceutical, plastics and chemical industries.

The 20th Century Legacy

Part II

Technology's Effect on Society

The 20th Century Legacy

4. Technology and the Economy

Accompanying the remarkable advances of technology since the Industrial Revolution has come economic growth and an improved standard of living. Economists assume that technology and innovation caused the economic growth and conclude that further advances in technology will lead to future prosperity.[4,5] Robert Gordon noted, however, that per capita GDP growth has steadily declined since it peaked in 1950, even as new, high tech electronic industries have become major factors in society.[4] He opined that some technologies may be more effective than others in stimulating economic growth and that economic growth may return to the slow growth rate that preceded the Industrial Revolution. Tyler Cowen suggested that "the low hanging fruit", had been picked and further progress will be more difficult.[5]

The creation of power and new materials combined with the scientific revelation of how nature works are the major technical developments of the Industrial Revolution. Together they enhanced the physical well-being and economic comfort of the nation. The later electronic industries, however, expanded the mental capabilities of people with computers, information storage and retrieval, automation, and artificial intelligence, but did little to improve physical well-being. It is unlikely discoveries as important as mastery of nature and expanded utilization of the environment in the 20th century can be matched again in the future for increasing the physical comfort of people.

The young 21st century has experienced a financial crisis and collapse of the housing market, static employment numbers and slow economic growth. Income disparity is said to have widened at the expense of the "middle class". The U.S economy has now recovered to prerecession levels, but a growing retired and under employed population persists. Innovation, technology, and science have not been a panacea for economic prosperity. Professor Gordon is probably correct; slow economic growth is likely to continue to be the norm.

Technology Effect on the Economy

The Industrial Revolution saw the western world economies shift from a land based agricultural structure to a capital based mining and manufacturing structure. Most of the population was made up of laborers; "manpower" was the unit of energy. Economic theories of Adam Smith, Karl Marx, and others were derived for such economies where unskilled workers were dependent on the economic control of first the landed gentry and then

Technology and the Economy

exploiting capitalists. In the 20th century new machines changed the structure of society again, further reducing the labor content of industrial production.

In the U.S. the Industrial Revolution of the 19th century coincided with expansion of the new nation. Colonists and immigrants poured over the Appalachian Mountains to settle on free land. The westward migration continued through the century lured by mineral riches and fertile land. Immigration swelled the numbers as people fled from the constraints of Europe seeking freedom in the new land.[5] Expansion of the railroad to the Pacific shore in the latter part of the century tied the huge nation together. Railroad, steel, coal, mining, finance, ranching, farming, petroleum, and shipping industries were born to produce and market the products throughout the nation and across the seas. Invention of machinery increased productivity, but it was the expansion of commerce and trade and creation of new materials that caused economic growth, not machines.

Prior to World War II, the mining, agriculture, construction, and manufacturing industries employed many millions of low and semiskilled workers. Technical innovation then not only reduced the labor needed to provide for basic needs but also created large, new service industries, including medical, financial, transportation, communication, and, entertainment. These new industries that now are among the largest components of the economy require better educated professionals and trained specialists. They are so pervasive today it is hard to realize that they did not exist as major economic identities one hundred years ago. In addition to these new service industries, employment in government and education also grew enormously. Economists had expected that these new industries would provide more jobs than those that were lost, but this has not been the case.

The spectacular electronic communication innovations of recent years – computers, internet, smart phones – have greater economic benefits for the foreign countries where they are manufactured, than for the U.S. where they are created. Automation has not been able to overcome the high labor costs of manufacturing electronic devices and components in the U.S. It appears that the innovations in communications have less economic impacts on society than prior innovations in energy generation and creation of new materials. New machines and new products improved the quality and comfort of life, but electronic devices mainly provide sensual benefits.

As the 21st century dawned, computers and the internet began to shift commerce away from local stores and offices, reducing need for clerks, and upsetting the sales tax base so important to state and local governments. Technology has also made possible the growth of mega banks and complex financial instruments that created the financial meltdown of past decades.

Technology and the Economy

The available workforce expanded as women were employed in all occupations. Many families now have more than one wage earner. Families headed by employed single mothers and fathers are common place. Immigrants from depressed areas in Latin America and Asia seeking better lives have expanded the work force further. The numbers are reduced somewhat by fewer teen age workers due to more restrictive child labor laws and rising minimum wages.

The belligerents of World War II rebuilt their economies with U.S. economic assistance. After the collapse of communism in the Cold War, most countries in Europe and Asia adopted capitalistic systems and increased their living standards by adopting technological innovations. World trade expanded as large commercial enterprises operated internationally. Transportation and communication advances brought the world together. It became more profitable for manufacturing companies to outsource labor intensive operations to lower labor cost countries and to move corporate headquarters to lower tax nations.

The U.S. economy usually recovered from recessions within a few years in the past and resumed rapid growth. The current economy differs in that growth is much slower and underemployment persists. The net economic effect of technology has been to reduce employment, increase human longevity, and skew incomes toward the very wealthy. It is probable that this trend will continue in the future.

Technology has greatly affected society, but economists are wrong in assuming that Technology was the cause of economic growth in the 19th and 20th centuries. Technology is analogous to a tool kit (Chapter 12). Tools have little intrinsic value; their value comes from how they are used. Power has economic value only when it is directed to commercial gain.

Technologies change the social structure and the value of labor, but commerce and trade drive economic growth. Society today has little in common with that of the 19th century.

Situation Today

The employed percentage of the population remains at historically low numbers, as many have ceased looking for jobs and the retired population grows. The basic problem is that population growth exceeds the growth of jobs, and jobs have become highly specialized.

Recently, the economy has stagnated and immigration has slowed. Many people rely on welfare payments and government assistance programs to maintain adequate living conditions. Fortunately, inflation remains low.

The country and the world are awash with debt and the availability of money. Wealth is represented by numbered accounts in diverse financial and government institutions. Cash and checks are disappearing as electronic transfers of funds between financial accounts proliferate. Management of financial assets around the world has created many of the world's richest people. The financial world is separate from the economy but financial crises affect everyone. What is good for markets and investors is not necessarily good for workers and families.

Economic and social issues are growing in magnitude. To spur job growth government has proposed increased investment in infrastructure projects, assistance to grow small businesses, and education to prepare more applicants for technical jobs. Recognizing the shortage of qualified applicants in the U.S. for specialized jobs in the expanding high tech industries, government has encouraged greater focus on science in schools. Colleges are responding with emphasis on Science, Technology, Math, and Engineering courses - the STEM Initiative. This should increase the technical competence of graduates which is important whether or not they are directly employed in technical jobs.

To help those in need, government provides several social programs including unemployment benefits, tax incentives, food stamps, medical care, and low interest mortgage loans. Some governments are raising minimum wages to insure all receive "living wages", and they are increasing taxes on the wealthy to better equalize the distribution of national wealth. But there is a growing body of thought that in a free society inequality increases as economies grow. "Inequality was to a large extent created by the success of modern economic growth." [6] So perhaps we need to learn to accommodate diversity in wealth instead of trying to combat it.

Unemployment benefits may be a disincentive to look for jobs, as is often asserted, but for many of the long-term and older unemployed workers there are no suitable jobs to be found.

In recent years, government central banks have reduced interest rates to near zero to prevent failure of overextended large financial institutions and to stimulate economic recovery by expanding the money supply. This was intended to encourage investment by employers to expand production and increase employment. The result has been ineffective, probably, because companies and investment bankers already had access to more cash than they could profitably invest; and consumer demand was not growing to warrant expansion of production capacity. Some companies decided their most profitable investment opportunity was to buy another company in a foreign country with borrowed money, reorganize, and relocate corporate headquarters abroad to avoid high U.S. corporate taxes. Many companies increased returns to their shareholders by buying their own stock or by increasing dividends, again with borrowed money. The stimulus money has gone largely into the banking system encouraging increased leverage in a society that is already highly leveraged. The U.S. economy in recent years has morphed into a finance dominated structure that has magnified the divergence of incomes. A small group of super rich (0.1% of earners) receive a disproportionate share of personal income (12%) and consist of executives in large financial institutions, investment companies, and multi-national corporations.[6]

It seems that the net effect of the Fed's stimuli has been fewer jobs and a class of super wealthy.

Remarks

Technology's economic effect is to increase efficiency, reduce labor content, increase specialization and complexity, and abet merging of international financial and commercial interests worldwide. Technology complicates government's ability to manage domestic commerce. The economy is no longer growing fast enough to fund the needs and wants of the nation; taxes cannot be increased enough to cover the desired expenditures. So governments sell bonds to cover the shortfall transferring obligations to future generations. National debt is now nearing $20 trillion and growing.

People today are critical of unequal distribution of incomes and wish to make it more equal by higher taxes on the wealthy and boosting wages of the less fortunate. These policies have been ineffective, as wealth flees the country and higher wages impede commerce. Extreme wealth comes from monopolistic control of commerce. Owners of railroads in the 19th century controlled shipments of commodities and became wealthy. Owners of coal and oil in the 20th century controlled energy and became wealthy. Today owners of communication channels (cable, satellite, internet, and cyberspace) and finance are the super-rich. The super wealthy now are funding projects to raise the standard of living in foreign countries, buying luxuries that are made abroad, or investing in speculative domestic projects that provide little economic value to the country. The wealthy "Robber Barons" of the 19th century and their families created foundations that recycled their riches within the country to benefit later generations.

Corporate tax policies should be changed to levels of other industrialized countries to eliminate incentives for U.S. companies to transfer operations to lower tax countries. Excessive taxation of the rich encourages tax avoidance responses and dissipation of wealth to nonproductive uses. Tax policy incentives should be established to encourage the wealthy to recycle their wealth back into the economy for the benefit of U.S. citizens.

Technology and the Economy

Minimum wage laws should be curtailed; they restrict the ability of small businesses to compete. Actions to better distribute incomes exacerbates the basic unemployment issue by encouraging employers to improve efficiencies through further cost cutting actions and to export low paying jobs to lower cost nations. A largely unrecognized deleterious consequence of the labor laws is the elimination of menial jobs for youth. In the past these humble, low paying jobs provided teens the opportunity to learn personal responsibility and moral values. Not all jobs need to provide living wages.

The conventional economic remedies of the past are not solving today's unemployment and income divergence problems because society and the economic structure have changed. In the past the economy largely made things that were labor intensive - food, clothing, materials, construction, and machinery. Technology reduced the labor needed to make things and created large new industries doing things that require special intellectual talents. It may be more difficult to manage the economics of doing things than that of making and selling things, particularly in a highly leveraged global economy. The wide availability of cheap money inflates real estate values making it difficult for low wage earners to get adequate housing.

The high cost of government is a great burden to commerce. There are too many regulations and laws that curtail business operations and increase costs. Accidents happen – we need not assign culpability and costs to people involved. Liability insurance is a costly burden for business, and ties up the courts. Class Action law suits primarily profit the lawyers.

New ideas are needed to reduce the oversupply of the working population by both creative job stimulus programs and more restrictive immigration policies. Large unemployed populations of young people create unstable societies, as now evident in the Middle East.

5. Technology and Government

The United States and the Industrial Revolution are closely linked. Both began about 1790 but did not reach maturity until the 1950s. Inventions of steam engines and remarkable new communication methods in Europe were quickly adopted in the U.S. to shrink the vast distances in North America. The Treaty of Paris in 1783, which ended the Revolutionary War, ceded to the Colonies all of the land from the Atlantic to the Mississippi. Twenty years later, Napoleon ceded France's large remaining interest in North America, to the United States - the "Louisiana Purchase". The vast sparsely populated domain of the new nation was comparable in size to all of Europe. The treaties did not, however, bring peace and the fledgling nation was tested by England again in 1812.

The nation was forged by wars: with England, with pirates on the Barbary Coast of Africa, with Spain and Mexico, and by continuous conflicts with the indigenous Indian tribes. The devastating civil war in the 1860s established the dominance of the Federal Government over the states. As the 20th century began the U.S. moved to the world stage with the conquest of the Spanish in Cuba and in the Philippine islands and, with mediation of the Russo - Japanese war in Asia. There followed two world wars, and years of hot and cold wars to prevent the spread of communism.

Few seem to appreciate that the U.S. we know today is really just one hundred years old. The last two of the forty-eight states comprising the continental U.S. were not added until 1912. The U.S. of the 20th century differs markedly from the frontier nation of the 19th century and from that of the writers of the Constitution. It is often said that the U.S. is a nation of immigrants seeking better lives. This is certainly true, but the immigrants of the 19th century were seeking freedom and building the nation, whereas, most immigrants in the 20th century fled poverty and conflicts at home to enjoy the benefits of U.S. prosperity while retaining allegiance to their home country. The U.S. is a nation of laws; illegal immigrants have never been welcome.

The U.S. was founded on a tradition of trade. In 1800 the U.S. was second in the world only to England in international trade - the U.S. merchant fleet traded throughout the Atlantic.[7] While seeking peace, the U.S. also has a tradition of war to protect its commerce, its citizens, and its allies. Soon after the Federal Government was inaugurated, the coast guard and the navy were created to protect the merchant ships. For the first century of its existence the U.S. relied on state militias and citizen volunteers in major conflicts; the standing Federal army was small and served mainly to protect settlers as they migrated westward. The right for citizens to bear arms was essential for personal protection in the frontiers as well as for participation in the state militias. Both Army and Navy have grown in size and competence in the 20th century as the nation became the leading world power. Over the years, weapons technology has been an important factor in the U.S. culture.

When established on the east coast of North America in the 18th century, each colony maintained a relationship with its European sponsor; there was little interchange among the colonies. They had different cultures and economic structures ranging from plantations in the southern colonies to small farms, religious settlements, and sea ports in New England. Only New York, founded by Dutch traders, traded freely with the other colonies. Freedom and independence were cherished, so when the English government imposed greater restrictions and demands, the colonies cooperated in resisting English rule.

After the revolutionary War was ended in 1779, the colonies formed a Confederation to create a common front against a hostile world. The new nation differed from those in Europe in greater size, in the wide dispersal of the population in small towns and hamlets, and in the crush of emigrants fleeing from Europe. Because of the great expanses of the nation, most government was local and the states resisted delegating the necessary powers and money to the central government.

There followed the "Second Revolution", ten years of debate between the Federalists advocating a central government and those wanting to retain the autonomy and freedoms of the individual states. Never the less, all recognized that the colonies could not long survive alone. The debate culminated in the Constitutional Convention in 1787 in which delegates from the states, composed of some of the greatest visionaries of government science, assembled to create a new government.

Constitutional Considerations [8, 13]

Most governments of the world at the time were aristocracies; government laws and justice were centered in the executive branch of government. Having fled from such governments, the colonists wanted to maintain their cultures and independence without interference by a strong federal executive. They wanted a government based on laws and popularly elected federal and state legislatures. Most government had to be local with the federal government restricted to foreign affairs and supporting commerce.

After four months of intense discussions, a constitution was designed that derived power from the people but coordinated the responsibilities of federal, state, and local governments through processes of delegation of responsibilities and checks and balances. The states were responsible for internal affairs and the federal government was to handle areas that could not be handled effectively by the states

The resulting Constitution provided for a federal government composed of independent legislative, executive, and judicial branches operating in concert. The House of Representatives, the legislative focal point of government, was the only branch elected directly by the people. The Senate's function was to monitor and approve laws passed by the House, and approve appointments and treaties made by the president. Senators were selected by vote of the state governments. The president and vice president were elected by the vote of "electors" selected specifically for that purpose by the state governments. Ten amendments (The Bill of Rights) were added to specify the generally understood, inalienable rights of individuals, before the Constitution was ratified by the thirteen states. This unique constitutional structure was designed to govern a large multi-cultural society and protect minority cultures from dominance by majorities.

Time and technology in the subsequent two hundred years have caused changes that have negated some of the founder's important intentions. First the Civil War in the 1860s, fought to prevent southern states from seceding from the union, established the dominance of the federal government over the states. Then in 1913 the Seventeenth Amendment provided that senators would be elected by popular ballot instead of selection by state legislators. This removed the important check the states had to prevent laws that adversely affect state rights. The current practice for the Electoral College to mimic the popular vote for president further reduced the protection of state's rights. Time has erased the checks in the Constitution to prevent the dictatorial dangers to personal freedoms posed by direct democratic government.

Technology Effect on Government

Technology advanced as the Nation grew and served to reduce the effects of distance on transportation and communication. In the 19th century railroads enhanced movement of freight and people across the growing expanses. Telegraph extended the reach of messages. Steam ships shortened travel times across the oceans. In the 20th century automobiles and telephones brought these advantages to individuals.

Originally, a newly elected Federal government was to start on March 4 of the year following national elections to allow those elected to relocate to the Federal Capital. This resulted in four-month "lame duck" governments between the times of election and inauguration. Evolving communication and transportation technologies over the next century speeded and expanded the electoral process so that in 1937 the 20th Amendment was ratified moving the inauguration up to January 20 for the President and up to January 3 for Congress.

Today's government institutions were made possible by the evolution of radio, movies, and television, and by development of rapid, long distance transportation. By 1940 radio and movies allowed the President to speak directly to citizens on special occasions from his office in Washington. With the advent of television, the President now appears in our living rooms and talks with us whenever he likes from wherever he may be. Electronics and the internet have further increased Government's control over the nation. The Federal government can now reach all citizens directly, increasing the power of the Federal Government over the states and individuals.

Candidates for President of both political parties now announce their intent on public media nearly two years before the actual election. They track their voter support by frequent polls of the electorate. They debate other candidates in nationally televised events in key venues around the country. Politicians in the competing political parties no longer select their best candidates to face off in the Presidential election: the party candidates themselves are now selected by popular vote. Television appearance, glib tongue, money, clever public relations, and skilled use of social media are necessary qualities for election. Political experience and management qualities no longer impress voters. More technical savvy but less experienced younger voters are becoming a bigger factor in the primary process.

Technology has affected the operation of Congress and the Federal bureaucracy by creating the need for lobbyists to explain the specialized needs and operations of the many conflicting interest factions to legislators and regulators. Without lobbyists government would be operating blindly. As the economy and nation have grown, lobbyists have become powerful forces in government, often supporting favored candidates for political office. Lobbyists for industries, trade unions, social organizations, government workers, and commercial organizations compete to influence legislators for favorable treatment. Those on the losing side of issues blame opposing lobbyists ("Special Interest Groups") of undue influence, but as long as legislators and regulators listen to both sides and do not accept bribes the system is necessary. The danger comes when lobbyists and regulators bypass the elected legislators.

Government Regulation of Technology

As technology has had a great effect on the structure and functions of our government; so also Government affects the progress of technology by regulations and by direct

Technology and Government

and indirect support of favored technical projects and programs.

Evolving technologies in the 20th century, such as electricity, automobiles, aircraft, use of air waves, food quality, and medicines created need for regulation at the same time that new technologies enhanced the capabilities of federal government to regulate.

The Constitution requires the Executive Branch of Federal Government to enforce the laws passed by Congress. So, hundreds of departments, agencies, bureaus, and commissions have been created to enforce the laws by imposing rules and expanding regulations. Today there are fifteen departments in the President's Cabinet compared to four (State, Treasury, War, and Postal) in George Washington's original cabinet. Departments of Justice, Interior, and Agriculture were added in the 19th century, and ten more were added since; the last being Homeland Security in 2002 after the terrorist attack on the World Trade Center in New York. In addition to the basic departments of Treasury, State, and Defense, and now Homeland Security, most others regulate specific areas of the economy: Commerce, Labor, Health & Welfare, Housing, Transportation, Energy, Education, and Veterans Affairs. These Departments are headed by officials appointed by the President. Also many agencies, bureaus, and commissions have been created within these departments by Congress to address specific issues. These government bodies employ millions of civil service workers.

New and rapidly developing technologies make it difficult for regulators to keep up with advances of those they regulate, and to recognize the unintended collateral economic and social impact of their decisions. Government regulators focus on intended benefits with little consideration of resultant costs. This problem for government regulators was recognized in the 1930s by a movement fostered by Howard Scott called "Technocracy" that advocated that government be administrated by specialists skilled in management with a solution mindset for better technical decision making. This vague idea never got very far with politicians and soon faded away.

Regulations are, in effect, laws imposed by the executive branch of government ("Administrative Law"). The Founding Fathers had been suffering from such laws imposed by the King of England and his appointed local governors. So they tried to prevent them in the Constitution by limiting the powers of the executive branch. Government today has taken a different approach that greatly increases the power of the President. The Federal Register, which lists all Federal regulations, encompasses over 79,000 pages with several thousand rules being added yearly by the many departments, agencies, and commissions. Regulation costs are nearly half of the Federal budget and cost society even more to comply. (Clyde Wayne Crews, Forbes, June 30, 2014).

It is the nature of organizations to perpetuate themselves, so government agencies often persist making rules after their initial purpose is no longer needed. An example is the Department of Energy that was created in 1977 to establish energy independence from foreign oil, then shifted objective to development of renewal energy processes to replace fossil fuels. Regulation can be open ended such that rules become increasingly restrictive, going beyond the intent of the originally legislated law. It is human nature to over react when there is neither specified objective nor end point. Environmental protection laws, for instance, have been applied to protection of obscure fish and animal species often at considerable economic costs to the public. Such laws do little to protect the environment from destructive practices.

The more rules a government makes, the larger it must grow to administer the expanded program. Some agencies grow too large, too costly, to fail; so government bureaucracy becomes uncontrollable. Through bureaucratic expansion Federal, State, and local Governments have grown to be a large part of the economy.

When companies grow so large that they dominate their industries, they can be broken up by anti-trust laws or subdivided for more efficient management. However, there is no practical cure for governments that grow too big to manage. The federal government and some state governments are now suffering from poor management of some functions.

Government Direction of Science & Technology

Prior to World War II science and technology were the purview of the private sector. During the war scientists, engineers, and executives from universities and industry were employed to assist the armed services to develop new weapons like the atom bomb. After the war the National Science Foundation was formed, followed by the National Aeronautical and Space Administration, and the National Institute of Health, to sponsor research in space exploration, nuclear physics, and medicine.

Each branch of the military also invests in programs to provide the high tech weapons and systems needed to fulfill its specific missions. Continual improvement in technology is necessary to maintain world superiority. Military needs can be a powerful stimuli for innovation that sometimes leads to new civilian applications.

Space and the Atom
The highlight of the NASA program came in 1969 with landing humans on the moon. NASA is continuing to explore the solar system and outer space with satellite telescopes and with space probes. Together with other na-

tions, the U.S. is conducting research in the orbiting International Space Station on the effect of space on humans, materials, and processes. Commercial interests are now trying to cash in on space opportunities by providing rockets to ferry supplies and astronauts to the Space Station and hoping to take tourists for rides in space. The exploration of space has provided exciting achievements that make Americans proud.

In recent years NASA expanded its mission with satellites to study Earth's surface, the weather, and the composition of the atmosphere. NASA has become a leading advocate for human cause of Global Warming. With other nations the U.S. also funds construction and operation of large high energy particle accelerators to explore subatomic matter.

Space exploration and atomic research are very expensive programs now aimed at learning more of the workings of our universe. The further scientists push into space and into the interior of atoms the greater the costs. People expect that fallout from these studies will prove useful, but actually they provide little economic value beyond the employment of throngs of scientists and engineers. Scientific research is a government luxury and should be monitored to prevent runaway expenditures.

Medical & Health

The National Institute of Health is the nation's medical research agency that fosters scientific research in medicine. It is composed of twenty-one institutes and several support centers focusing on particular diseases and body parts. It provides research grants to scientists in many universities and hospitals. These agencies monitor the health of the nation and approve and recommend pharmaceuticals, medical devises, and health practices.

Energy & Ecology

Government has taken a very active regulatory and financial role in directing Energy and Ecology programs. Chapters 2 & 8 address these complex issues.

Remarks

Government & Scientific Research

Government is poorly structured to direct scientific research and technology. Government officials excel in getting elected, a talent that requires different skills from those needed to run a business or manage investments. Government is not goal oriented, except in time of war; it is focused on the present and subject to changing political and electoral pressures. There is no mechanism to define goals, timing, costs, end points or the effect on society. Government is not equipped to prioritize activities; it follows the preferences of powerful political lobbies and electoral majorities. Programs change direction as politics change. Government administrators are not accountable for the costs and results of their programs, which often affect subsequent generations long after their terms in office, expire. Political benefits to supporters often top concern for the public.

Government is complex with many over lapping, independent federal and state components that are each affected by different constituent interests. Large technology programs impact many areas of society which are the concern of various autonomous and poorly coordinated federal and state governmental agencies. Each agency is motivated to perpetuate itself and to grow so focus is on expanding budget and selling the benefits it provides.

Government is not structured to earn money for the citizenry, only to take money by taxation or regulatory fees. Large government technical programs are undertaken to gather information to support a political objective with little consideration of economic impact. The large Energy and Ecology programs discussed in Chapters 2 and 10 are examples of pushing programs ahead toward wishful objectives with fuzzy visions of social and economic consequences, and without allowances for acquiring the necessary supporting technologies and financial backing.

Support of medical research by National Institute of Health is appropriate; but how far in space need we explore? Is money better spent on space or nuclear research than on social programs or infrastructure projects if all cannot be funded within a balanced budget? Has NASA grown too large to fail?

On leaving office in 1961 President Eisenhower warned of the dangers of the concentration of power, in a military-industrial complex, and in a permanent armaments industry. Perhaps he was thinking of Hitler's close association with the I. G. Farben Industries, a large industrial complex that powered the Nazi war machine in World War II. Such a concentration of power has not occurred. However, Eisenhower also warned of the potential danger from the formation of a powerful network of government funded experts and scientific-technological elite, assuming undue influence. The growing influence of NASA and government funding of pet research projects in universities comprise such a danger.

Robert Piccioni explained why government should fund Science? [9] "Why do we devote so much time and money to music, art, and sports? Because they enrich the human soul and they make life exciting and enjoyable." Just as climbing a mountain to see what is on the other side or spending time discovering the mysteries of nature satisfies our innate curiosity. To improve our lives, we must invest in science."

Government funding of science and the arts is appropriate and desirable if it can be funded with discretionary dollars after essential government obligations are fulfilled. Recipients should be free to pursue their ideas with government oversight but not government direction. Because few understand the workings of science, however, government control of science is a peril to be avoided.

Government & Technology
The U.S. was founded on the principle of individual rights and freedom of opportunity that allowed for commerce and foreign trade. Government's role was to protect those rights. Americans were quick to defend their freedoms and assist other peoples to achieve theirs. Although the U.S. has been the leading world economic power for over a century, many U.S. citizens are isolationists and want to hide behind our two protective oceans. Technology today makes this impossible. If we do not protect our worldwide commercial interests, other powers will take them from us and threaten our freedom.

The Constitution defined a representative democracy, a republic that retained the principle advantages of democracy and of a parliamentary autocracy while eliminating their principal deficiencies. This proved ideal for the new, enormous but sparsely populated, multi-cultural United States. Now technology is taking the U.S. away from the intended republican representative democracy to the direct popular election of the President. Only the formality of the states' Electoral College and the moderating power of political parties stand in the way of direct popular election of the President. The protections for freedom and independence of the people that were designed into the Constitution have been compromised.

The electorate now is split between those advocating free enterprise with the private sector dealing with social issues (Republicans / capitalists) and those advocating government dealing with all issues (Democrats / socialists). The contesting political factions vociferously advocate their differing views for the future and how to get there but there is little agreement on the basic economic and social facts. The conservative extreme of the Republican party advocate elimination of deficit spending knowing continued deficits will ultimately lead to financial collapse. Their proposed solutions to rapidly eliminate deficits, however, would cause considerable social distress in the near term. On the other hand, the socialist extreme of the Democratic Party argues for expanded social programs, expecting the wealthiest people and corporations to fund the programs, even though the costs of benefits exceed the capabilities of the rich to pay. So the national debt continues to grow.

Since the world and national issues are much different today, it is understandable that government structure would change from that which the founder's created. Federal government now dominates over the states, and administrative laws formed by the Executive Branch and the Supreme Court diminish the power of the Congress.

Technology and Government

Although the founders intended a minimal federal government, it has become a major employer with major effect on the economy.

Later in Chapter 11 it is noted that organizations grow like living organisms changing purpose over time, but well run organizations have clear mission statements to guide them. The basic purpose of government should be to protect citizens from harm by outsiders and from each other. Formerly, governments were focused on insuring the purity and safe handling of food and water supplies for all. Now governments are concerned with the physical well-being of individuals, protecting citizens from themselves. Governments try to curb unhealthy practices, such as use of drugs, tobacco, fatty foods, salt, and sugar. They promote serving only "healthy" foods in public institutions. The Occupational Safety and Health Administration define and regulate safe operating practices in fine detail for factory workers and operators of machinery. Accidents are often followed by new regulations to prevent recurrence. Each regulation diminishes personal freedom. Governments have become much more intrusive in our lives. Personal health and safety are credible issues, but should government be so intrusive? Government should act like a leash preventing people from excessive behavior, not as reins to restrict their freedom to act.

The prosperity and complexity that technology has brought to society also has confused and diverted our values. Before Technology personal freedom of opportunity was the basic value; today conformance to popular social behavior is paramount. "Proper Speak" and cultural diversity have replaced racial integration and equal opportunity as social goals.

Some one hundred and twenty years ago, Alexander Tyler, a history professor at the University of Edinburgh, said "a democracy is always temporary ----- a democracy will continue to exist up until the time that voters discover that they can vote themselves generous gifts from the public treasury. From that moment on, the majority always votes for the candidates who promise the most benefits ...... with the result that every democracy will finally collapse over loose fiscal policy, always followed by a dictatorship." If Tyler's ideas hold, the U.S. is quite ill and is in dire need of a good doctor!

The writers of the Constitution, recognizing the failings of human behavior, designed a government of checks and balances to prevent the instability inherent in a democracy. The president was elected by delegates in the legislatures of the several states, not directly by the people. The federal government was checked by the states. The chaotic 2016 presidential election illustrates why the democratic election of the president leads to unstable government. This election shows how a crusader outside of government with a popular message can amass a huge following; think of the damage to democracy if that crusader had a military following. The shift of power away from the legislature in California by powerful lobbyists using voter propositions to enact laws is another threat to democracy. People seem not to have understood the importance of delegation of power in a democratic government. In our highly specialized technical society, the citizenry lacks the information necessary to make wise judgements of national issues and policies. It is necessary that all economic and social factions be represented in choosing the executive, but not the interests of every citizen.

With the popular election of both houses of Congress and the President (with the expectation that the Electoral College will mimic the popular vote), the U.S. is becoming a true democracy. The authors of the Constitution considered democracy to be an unstable form of government and had established checks to prevent this from happening.

California, a state that has always marched to its own drummer, has taken democracy a step further by enacting revenue laws by popular vote, marginalizing the importance of the legislature. It is instructive to consider what has happened to California, which is such an important component of the nation. In many ways California is emblematic of technology in the 20th century.

6. California Democracy

California, like Texas earlier, was extracted from Mexico by venturesome settlers and formed its own government before annexation by the Federal government and becoming a state in 1850. Unlike Texas, however, gold was discovered in California, and people rushed to the state expecting to find wealth lying in the ground. Some early settlers soon gained control of the land and mineral resources, and became rich selling land to those who came later. Settlers found that riches from gold could be very elusive.

The natural beauty, sunny climate, and moderate temperate have long attracted visitors and immigrants; people came for the beauty and many decided to stay. Immigrants fleeing from the plains states during the Dust Bowl of the 1930s built an agricultural industry that became famous nationally. A system of dams, canals, and reservoirs were developed to distribute water from the wet north to the fertile south. At one time, Los Angeles County led the nation in value of agricultural production, mainly citrus. Subsequently, as cities replaced farms in Southern California, counties in the San Joaquin Valley took leadership with orchards, vineyards, cotton fields, dairies, and livestock. Minerals, timber, and petroleum were also major factors in the growth of California. Automobiles and airplanes were features of California living since the 1920s, and commercial aircraft manufacture became a big industry.

In the years following World War II, many veterans who were introduced to California during the war returned. Now one in eight Americans lives in California. Land values have soared, and water is precious. The cost of living has become the highest in the nation; high taxes and onerous regulations have driven most manufacturing industries out of the state; aircraft manufacture was one of the first to leave in the 1970s. California has become the worst state in the country for industry and business.

Richard Rider, in an article entitled "Unaffordable California -- It doesn't have to be this way" (California Policy Center.Org, December 23, 2016), compared the tax, regulation, economic, and social policies of California with those of the rest of the nation. California ranks worst, or nearly so, in all categories. California has highest taxes, most regulations, highest Workers Compensation rates, lowest bond ratings, highest utility and transportation costs, highest unemployment, highest poverty, and poorest public school student achievement. California is the worst state to retire, and has been losing net population in recent years. Tragically, California led the nation in most categories sixty years ago.

How did things go so wrong after such a promising beginning? There must be many causes, but it is clear that government has failed its citizens, and shows no signs of improvement.

California Government

The California constitution provides for the traditional U.S. bicameral legislature, a governor, and courts; but adds a provision for popular recalls of elected officials and for popular enactment of laws and constitutional amendments.

The State government has become quite large; there are several hundred administrative agencies, boards, commissions, and departments. Once created few, if any, are ever terminated. Government is the largest employer in the State. and government workers are the largest voting bloc. For the most part compensation and benefits are generous in comparison to the private sector. Unfunded pension liabilities are a heavy burden for State and local administrators. Health and welfare payments for the indigent and unemployed are substantial.

Rapid inflation of property values in the 1970s caused residential property taxes to soar, creating financial distress for long time home owners. State government expenditures had grown higher than the average of the states. Government workers had become 15% of the state work force, twice that of 1950. Showing their displeasure with the growing government, the voters in 1978 passed Proposition 13, restricting the rate of property tax increases for homeowners. The resultant law also required a super 2/3 majority vote in the legislature for any new taxes, a hurdle that has been very hard to achieve. Since then State and local governments have had to rely on creative ways to increase revenues. Examples include refundable deposits on bottles, few of which are actually redeemed, and "Cap and Trade" regulations that require businesses to buy permits for carbon emissions. State government also has passed some expenses on to local governments, recently including moving convicts from overcrowded state prisons to local jails to comply with court orders. Some prisoners convicted of nonviolent crimes were freed before completing their sentences. Apparently, California either needs more prisons to house convicted criminals or fewer criminals. Perhaps California has too many laws requiring confinement for violators, or perhaps a new form of punishment should be created other than fines and incarceration.

Legislative term limits were imposed to enhance turnover of elected officials. Independent voter commissions have been created to revise election districts, formerly a legislative responsibility, and to monitor government performance. The two party political System was compromised by creating open primaries in 2012 permitting voters to cross party lines to vote in primary elections. The effect has been to reduce the power of the legislature and to increase the dominance of urban majority interests over regional minority interests. The legislature has been marginalized to dealing primarily with social issues.

Because of these changes legislators focus more attention on pleasing their constituents and working toward their next job. Over one thousand new laws are being generated annually to add to more than forty thousand already extant. Since it is hard to increase taxes for new endeavors, most of the many new laws enacted each year are directed to regulating behavior, and preventing the recurrence of accidents or ecological damage. Every law reduces somebody's freedom.

California like other states sells bonds to fund infrastructure improvements; in California bonds must be approved by direct vote of the electorate. With the taxation constraints on the legislature imposed by Proposition 13, bonds are sometimes used to fund operating expenses as well as facilities improvements. An example is Proposition 71 passed in 2004 to sell $3 billion of general obligation bonds over ten years to fund stem cell research and build research facilities in California universities. Voters were told profits from the research would pay back the costs. After twelve years, profits have been scarce to none so the bonds have the effect of an added taxpayers' expense.

California has become a strong advocate for environmental protection, discouraging all but "clean" industries in the State by restrictive permitting processes and regulations. The mining, agriculture, and energy companies that brought prosperity to the California, are hard pressed to adapt to the growing restrictions as the population over whelms the land and water resources.

Gasoline taxes designed to fund the cost of the extensive highway system are now falling short of expanding needs because of reduced consumption of gasoline by todays motor vehicles. The Legislative Analyst reported to the legislature that the costs of the "Cap and Trade" regulations were incorporated in the price of gasoline, raising the price 11 cents per gallon and costing drivers $2 billion per year (Ventura Star, April 8, 2016). This is in addition to a direct California gasoline tax of 12 cents and 8% sales tax making California gas the most expensive in the nation. A large portion of the money is dedicated to California's high speed rail project and other transportation projects, probably the real purpose of this "hidden tax".

The California public school system that once was the envy of the nation now struggles to achieve acceptable student performance in elementary and high schools. The state universities and colleges, that once were nearly free to Californians, have become so costly to students that many students are saddled with debt upon graduation. Students from out of state, many from other countries, often get preference on admission because they pay higher tuition. As the colleges struggle to meet admission diversity goals, many worthy applicants are rejected.

Every election, voters are faced with ten or more state propositions that few are able to understand. So voters largely following the recommendations of advertisers, public officials, and the media. Government is now run largely by wealthy lobbyists and a government bureaucracy supported by a poorly informed electorate. Californians, with a governmental system that is close to becoming a direct democracy, like to think they lead the nation in progress.

In the national elections, in November, 2016, the Democratic Party achieved a two thirds super majority in both houses of the state legislature. Combined with a democratic governor and two democratic senators, minority views no longer have any voice in California politics. Power is centered now in a governor directed bureaucracy. Unfettered democracy is leading California toward becoming an autocracy with the governor wielding absolute power.

Remarks

For many people California appears as a Garden of Eden, a land of beauty and sunshine, a land of the future, promising spectacular accomplishments. It is the land of movies, fairytale creatures, Disneyland, and space travel. It is a land of technology, artificial intelligence, virtual reality. It is not a land, however, of concern for economics nor of respect for property rights.

The California government promises benefits to citizens without adequate resources to fund them. Government expects the wealthy few or future generations to provide the money, but establishes policies that impede wealth generation. Californians need to wake up to the fact that there is no "pot of gold" to be discovered to pay for their pleasures and desires.

Clearly, a government is failing its citizens when it cannot build enough prisons to house convicted criminals, but yet seeks to force use of renewable energy without concern for the added costs to society, and embarks on a multibillion dollar project to build a high speed rail system to connect distant cities many years in the future.

7. Technology and Security

The 20th century saw widespread wars involving most people on Earth at one time or another. Most of the conflicts occurred among nations in Europe, Asia, and Africa; the conflicts in the Americas were mainly internal regime conflicts. For many years the U.S. was protected by its two oceans and the principle of the Monroe Doctrine (1823). So the wars of the world had little effect on the U.S. homeland, and Americans could feel comfortable being isolationists. Now, however, the oceans provide little protection because of the advances in technology, and the U.S. can no longer remain aloof from conflicts over seas. Powerful weapons that were created by the contending powers during past wars have spread throughout the world and now threaten world peace and domestic tranquility.

As the 19th century ended, the leading world powers were engaged in an arms race to produce powerful cannons and huge battle ships. The First World War introduced trench warfare, barbed wire, motor vehicles, armored tanks, submarines, reconnaissance balloons, airplanes, and poisonous chlorine gas. In the twenty years between the end of World War I and the start of World War II the performance of engine-powered war machines was greatly advanced. The Second World War began with a Blitzkrieg of tanks and airplanes that made fortified defenses obsolete and led to dispersed "fox-hole" protections for combatants. Weapons grew rapidly in power, speed, and range, extending the battlefields to entire nations and their inhabitants. Aircraft carriers, transporting fleets of war planes over the oceans, replaced battle ships as the core of the Navy. World War II ended with the creation of nuclear weapons capable of destroying all life on earth, and fearful people began to build hardened bomb shelters.

During the Cold War between NATO and the Soviet Union, supersonic air craft, rockets, and atomic powered submarines were developed that could deliver atomic weapons anywhere. Since that time those having atomic weapons and long range missiles have been trying to prevent their spread to other nations. The necessary knowledge to make nuclear weapons is available to all. Manufacture is difficult, however, requiring materials that are in restricted supply and the use of specialized production machinery. Control of nuclear technology to prevent misuse that would cause unthinkable destruction of life will forever remain a challenge. Deadly gases, microbes, and plant toxins have been developed also that can create havoc on civilian populations. Their use has been banned by the world powers, but they remain a threat in the hands of terrorists and rogue nations.

As the 20th century drew to a close, international conflicts subsided to be succeeded by "Police Actions" and cultural clashes in the Muslim nations of the Middle East.

Personal Weapons

The creation of personal weapons of great destructive power (automatic rifles, rockets, and plastic explosives) has brought chaos to the Middle-East where Muslim tribal factions struggle for dominance and influence among themselves and for the elimination of Israel. The conflict is waged by attacks on civilian targets by suicidal religious terrorists and by wholesale annihilation of prisoners. Mass violence has spread to the U.S. as access to automatic hand guns and explosives in the hands of criminals and people with grudges and personal issues threatens innocent bystanders. Powerful lasers can cause considerable damage to people and electronic devices when misused. Many in the U.S., hoping to keep weapons away from bad guys, want to ban access to personal weapons and restrict availability to law enforcement agencies. Personal weapons are readily available, however, from many sources around the world, so banning is unlikely to keep them away from those who want them, just as banning has been impossible to prevent the wide spread availability of illegal drugs.

Technology & Security

The Department of Homeland Security was formed in 2002 combining several government policing organizations that protect our borders. All airline passengers are now screened by federal agents before boarding airliners and secret police agents may be included among the passengers. These actions have prevented several attempts to destroy airliners, but at the expense of reduced personal freedom and comfort of millions of passengers. Border security patrol has been tightened, and all law enforcement agencies have increased alertness to terrorist threats at large assembles of people. Terrorism by lone bombers remains a threat to ships, trains, large gatherings of people, and public utilities.

Modern society is highly dependent on electricity and the telephone. Portable electronic devices rely on batteries that require frequent recharging and proximity to electric power sources. So everything goes dark if the power grid is interrupted for just a few days. Telecommunication is interrupted if there is damage to phone lines, Wi-Fi transmission towers, or communication satellites. Modern society cannot function without the continuous availability of electricity and telecommunication service. Protection of the power and telecommunication grids from damage by natural disasters and attacks by terrorists is a major challenge. Even the communication satellites are no longer safe from attack, as growing numbers of countries have long range rockets.

Drones & Robots

Unmanned vehicles of all sizes promise wide spread utility; radio controlled drones take cameras to places inaccessible or too dangerous for people. Ground to air rockets, lasers and even small drone aircraft have become threats to airliners in the U.S. and abroad. Drones toting cameras threaten the privacy of individuals and the security of facilities. Drones able to deliver packages also could expand the reach of terrorists. Today anyone can access and fly radio controlled drones. So far drones pose no security threat, but for how long? The nation that has long had regulations governing manned aircraft flight is struggling now with rules for flying drones and operating robots.

Cyber Space

The 20th century bequeathed an insidious new issue on mankind: the use and spread of information on the internet. Cell phones and cameras document daily activities and even control remote electronic devices for good or evil. Social media spread personal information across the internet. All intrude on personal consciousness. People now spend much of their time in an imaginary, virtual world.

The remarkable electronic developments of recent years have changed the world, but they have also created many challenges that threaten the functioning of institutions

Technology & Security

and the freedom of individuals. The cyber world recognizes no geographical boundaries and threatens established national, political, cultural, and commercial systems. Corporate and government internal systems are vulnerable to exposure of trade secrets by cyber spies.

All personal devices connected to the internet are subject to infiltration by thieves seeking access to personal financial accounts and identity information. The wide spread availability of tiny digital cameras in cell phones and the ease of digital manipulation of images can adversely impact the personal lives of individuals. Social networks spread personal information widely that can be used for nefarious purposes. Hackers install malware on personal computers and smart phones to create havoc and to access personal financial accounts.

Governments have access to personal information and records on everyone through the Internal Revenue Service, Social Security, and Medicare systems. Geographical Positioning Satellites (GPS) and surveillance photography can track individuals and record their activities.

Broadcast of crimes and police actions quickly spread nationally and beyond through television and social media such that the accused are tried by public opinion before the legal system can act. Reports sometimes incite mob demonstrations that impede the course of justice. Employers may be pressured by public opinion to discipline accused for bad behavior, whether or not a punishable crime was committed. Personal information of individuals is sometimes retrieved and publicized to compromise their reputations. Humans seem to get perverse enjoyment from hearing about the sins and misdeed of others and judging appropriate punishments for the miscreants. Cyber space can be brutal to individuals and their personal freedom; it is particularly damaging to children and teens that lack the experience to understand and judge the relevance of what they see.

The access and control of personal information by government combined with governmental control of weapons creates a threat to individual freedom and to the preservation of our democracy.

Economy

By improving economic efficiency, Technology has created large numbers of unemployed youth with idle time and little responsibility. They are bound together and influenced by social networks and drawn to protest presumed social injustices. This large and easily influenced young population poses a great threat to the stability of society. A working population that exceeds the jobs available is unhealthy for society. Developing responsible occupations for all the population, skilled and unskilled, is a critical issue.

Spending outlays to support government operations, defense, and social programs, exceeds tax income. Economic growth has slowed, and deficits are rising. The Congressional Budget Office recently reported "The federal budget outlook is projected to worsen considerably over the next three decades under current law, with debt growing larger in relation to the economy than ever recorded in U.S. history."[14] Without a change in economic policies, economic collapse is not far away. Yet the political leaders and the electorate seem unable to make the necessary sacrifices now to avoid disaster. Deferral of such a growing problem only makes the consequences worse. Economic instability may be the greatest danger to the U.S. experience with democracy.

Remarks

Americans have a long tradition of foreign commerce and travel to all parts of the world. The U.S. navy and marines were always close at hand to protect them, and conflicts were kept far from the home shores. International trade remains an important commercial interest for the U.S. and its trading partners around the world. Protection requires maintaining a strong and steady military, diplomatic, and intelligence presence. The vacillation of foreign policy and military presence with changing political administrations confuses friends and encourages foes, leading to instability. We need to improve the consistency of foreign policy with defined limits we will defend that all will know.

Technology has magnified the reach and effect of terrorism with lethal weapons and cyber communications. National borders are easily breached spreading attacks throughout the world. Fear chases families from their homes to resettlement camps which become breeding grounds for more terrorists. Large emigration of refugees to foreign lands creates major economic and cultural issues for the host nation. Islamic terrorism has become a cancer that is spreading from Asia to Europe.

The greatest challenge the U.S. faces in coming years is protection of our independence, personal freedom, property, and privacy in a contentious world. The U.S. experiment in democracy has survived and grown over two centuries as the U.S. has become the most powerful nation in the world. Like other powerful predecessors in the past, it is facing major challenges to its dominance and even survival.

The world has always been a dangerous place, but the 20th century was particularly destructive and brutal. Conflict, war, and genocide occupied most of the 20th century killing hundreds of millions of people around the globe. The U.S. is hard pressed to protect its citizens and commerce and to support allies in unsettled regions.

In his studies of the wars of the 20th century, Niall Ferguson concluded that genocidal regimes and policies arise in the wake of declining empires creating power vacuums. "All people needed to do were to identify a group of their fellow men as aliens and then kill them. To avoid another century of conflict we must understand the dark forces that conjure up ethnic conflict out of economic crises. They are forces that stir within us still." [15]

Part III

Ecology

8. Pollution

A natural practice of people is to throw away what is not needed and bury or dump it on nearby land or water. Growing urban populations and rapid economic growth after WWII created massive water and air pollution problems in the nation. Waste from mining operations created ugly desecration of surrounding land and contamination of watersheds. Disposal of manufacturing waste and used consumer products filled trash dumps and contaminated land and water. Runoff of agricultural fertilizers and pest control products polluted nearby watersheds and caused unwanted aquatic plant growth and harmed fish. The arid west land, administered by the Bureau of Land Management, was desecrated by strip mining and by over-grazing of cattle. Old growth forests were decimated for construction lumber and furniture hard woods. It was common practice for cities and industries to dump waste directly in the nation's waterways without treatment; these were the same waterways that provided potable water for downstream cities and farms. Factories and petroleum refineries spewed waste gases into the air. Before the availability of natural gas, coal furnaces and stoves spewed smoke that caused ugly air pollution in many cities. Widespread shift to gas heating eliminated this problem.

Following World War II the green movement brought attention to the desecration of natural resources and pollution by the wasteful practices of the industrial world. The green advocates spurred governments in Europe and the U.S. to take action. Factories and municipalities undertook broad programs to remove pollution from wastewater. Then in 1962 Rachel Carson, in her book "Silent Spring" brought to public attention the damage to the environment of widespread spraying of pesticides, including DDT. This served to expand and accelerate the green movement.

Reacting to the destructive practices exposed by environmentalists, the Environmental Protection Agency (EPA) was created in 1970 to provide government standards for regulation of water and air quality. The same year the Natural Resources Defense Council (NRDC) was formed by a group of lawyers to agitate for enactment of environmental protection laws and for more effective enforcement of the current laws. Led by founder John H. Adams, NRDC has been at the forefront of environmental protection, and in 2010 numbered 1.2 million members.[10]

Pollution

The green movement, led by the legal challenges of the NRDC, has been highly successful in curtailing pollution and wasteful destruction of resources in the U.S., although many of the developing countries feel less constrained as they adopt the benefits of technology. After successfully curtailing the most egregious practices, the movement has now expanded focus to restoring nature to the way it was before human desecration. The stated mission of NRDC is "to safeguard the planet: its people, its plants and animals, and the natural systems on which all life depends." [10] This includes establishing national monuments to protect wildlife on land and sea, reintroducing wild animals to national forests, and restoring streams and rivers for spawning fish.

Water Quality

Many of the most egregious industrial and municipal water pollution practices now have been corrected in the U.S. Today municipal wastewater often is treated to the standard of potable water by sewage plants. Most industrial operations also have instituted processes for treating waste effluents to minimize damage to waterways. Considerable progress has been made, but there remains much to be done to further reduce environmental damage. Pollutants enter water ways by drainage from farm lands and run off from storm drains. Small amounts of hazardous materials sometimes still get through, and waste retention ponds in mining and manufacturing operations sometimes leak into adjacent waterways. Storage of hazardous waste is a growing problem.

People have long considered the oceans so vast that no harm would come from discarding trash from passing ships and from coastal cities. As prosperity and the benefits of technology have spread around the world, commerce and international trade have expanded across the oceans. Thousands of cargo ships, oil tankers, fishing boats, and ocean liners cross the oceans and clog the many ports. Millions of small pleasure boats ply the coastal waters and marinas. Trash tossed on beaches and washed down rivers and urban storm drains contaminates beaches and seaports. Fuel leaks from ships and oil spills from drilling operations and ship accidents create damage to sea life and to sea coasts.

Biodegradable materials may create health issues along the shores, but they are quickly dispersed, naturally, in the open sea. Durable plastic, metal, and glass packaging materials and other plastic products have become big problems because they degrade very slowly, and can be toxic to marine life. Local ordinances to limit use of plastic bags and to recycle plastic bottles have had only limited success reducing the quantities reaching the oceans. Lost durable plastic fishing lines and nets are hazardous to marine life, particularly dolphins and whales. Oil released accidentally from ships and drilling platforms cause con-

Pollution

siderable damage to marine life and sea coasts, but effects are soon healed.

Air Quality

The Los Angeles basin, periodically, is blanketed by stagnant air. A temperature inversion layer forms wherein the air at ground level is cooler than the air several thousand feet above. Because of the surrounding mountains and lack of wind, smoke and other pollutants settle close to the ground creating a health hazard. In 1967 California established the Air Resources Board to address this growing air pollution problem. A project was established at Stanford Research Institute to identify the composition of smog and how it is formed. It was soon shown that exhausts from motor vehicles and solvent effluents from industrial plants were the cause. Ozone, a strong eye irritant, was formed by reaction of volatile organic compounds (VOC) and sunlight. In addition to water vapor and carbon dioxide, which are the end products of complete fuel combustion, vehicle exhausts contained (VOC) from incomplete fuel combustion as well as oxides of nitrogen caused by the high temperatures and pressures of the automobile engines. There was also lead from anti-knock additives in gasoline and dust along highways from tire wear. Diesel powered trucks and busses spewed black, sooty exhaust and corrosive sulfur dioxide gasses. Pollution was particularly repulsive in areas around urban transportation centers.

When the Federal Environmental Protection Agency was established in 1970 and sought to impose regulations, the Air Resources Board resisted since they already had regulations in place and knew better what was needed. The EPA relented and California is still the only state allowed such a regulatory agency.

The automobile companies and fuel providers initiated broad programs to eliminate these sources of air pollution. High compression automobile engines, that had been optimized previously to provide high performance, were replaced by lower compression engines to eliminate nitrogen oxides. Catalytic converters were added to all cars to eliminate contaminants from incomplete combustion. Gasoline was reformulated to eliminate lead, and tires were developed for reduced wear. Gasoline pumps in filling stations were modified to recycle the displaced gas vapors during refueling. Fuel economy was improved further by reducing the weight of vehicles through use of aluminum, plastic, and composite components and later by electronic monitors and computer controls. Diesel engines were redesigned, and low sulfur diesel fuels were developed to meet EPA standards. Most old gas guzzling vehicles now have been retired. These process changes greatly reduced particulates and ozone formation in the atmosphere and reduced smog to manageable levels.

Industrial manufacturing plants also have nearly eliminated VOC emissions first by incinerating solvent effluents

Pollution

and then by adopting solvent-free processes wherever possible. In the 1970s and 1980s, hot melt, powder coating, and aqueous emulsion and electro-coating processes were developed to replace solvents in household paints, industrial coatings, adhesives, and inks. Solvent effluents have now been largely eliminated from the air. These actions have reduced smog in Los Angeles to levels that now cause little inconvenience.

Much of the central California valleys also suffer from stagnant air. Pollution generated by the growing farming and industrial activity in the valleys augmented by pollution blown in from the San Francisco Bay region has become an issue. In this case the pollution is mostly dust and smoke.

Some thirty years ago studies suggested that long-lived, halogenated solvent effluents from sprayer propellants, coolants, and dry cleaning establishments might destroy the ozone layer in the upper atmosphere that protects earth from ultraviolet radiation damage. It was discovered later that there was an "ozone hole" over Antarctica, which environmentalists concluded must be caused by halogenated solvent emissions and warned of a growing threat of radiation exposure to people in the southern hemisphere. Governments required that such halogenated products be replaced totally by other materials, and less effective but acceptable alternatives were developed. In subsequent years it was found that the ozone hole varies seasonally over time and is not an increasing threat to humans. More recent chemical studies suggest that the effect of halogenated solvents may be less than previously thought and that there may be other causes of the periodic ozone depletion phenomenon over Antarctica.

Post-Consumer Waste

Technology has reduced the costs of many consumer products so that it is more convenient and cheaper to discard used products than to repair them. Scrap appliances, vehicles, and electronic devices which are composed of many different plastic, metallic, and ceramic parts present special disposal problems. Most packaging materials are disposable as well. Trash accumulates in cities, along waterways, and wherever people congregate.

Some governments now require deposits on beverage bottles to encourage recycling, and some have banned use of disposable plastic grocery bags. Recycling of waste is preferable, but it is complicated by the difficulty of collecting and sorting trash. Many kinds of metals, glass, and plastics compose the non-biodegradable materials in waste. Collection, sorting, and disposal of solid waste is a growing problem as land burial sites are exhausted. Developing better technologies for disposal of solid waste will remain a continuing challenge.

Pollution

Hazardous Waste

Some byproducts of energy, chemical, and electronic technologies are toxic materials. Nuclear power plants create highly radioactive waste that must be discarded in ways to prevent contact with humans and other living things for centuries. Search for a suitable underground site in the U.S. is still debated since no one wants to have such waste sites nearby. Waste disposal remains a critical issue for nuclear power reactors.

Effluents from many manufacturing and processing factories contain minor amounts of toxic chemical, metallic, and other residues that create disposal issues. Waste from mining operations often contains toxic components that require special handling to prevent contamination of water sheds. Coal mining is a major producer of waste containing hazardous components.

Post-consumer waste is complicated by disposal of the many new high tech devices. Small batteries from electronic devices and large electrical storage batteries require special disposal, as do discarded light bulbs. Pharmaceuticals and their metabolic derivatives contaminate sewage disposal plants. In the past common building materials contained components that are now considered hazardous, such as lead, arsenic, mercury, and asbestos. So, great care is exercised in remodeling and razing old buildings. Proper disposal and recycling of industrial, consumer, medical, and electronic waste are growing in importance with the expansion of technology. Hazardous waste is a legacy of technology that requires continuous management attention.

"Species" Contamination

Local environments have always been subject to change as weather, ocean currents, and species migration has carried plant, animal, and microbes to new venues. Bubonic plague spread across Asia and throughout Europe in the Middle-Ages. Change of environmental conditions as well as destruction and evolution of plant and animal species are eternal natural phenomena. However, species contamination is much faster today because of the much greater worldwide travel and commerce. International travel and transport of goods often carry along germs, insect, plant, and animal species that take up unwanted residence in new locations. These foreign species compete with local counterparts often causing great ecological damage. Citizens on occasion have purposely introduced plants and animals that proliferate in the new environment. Notable examples include gorse from Scotland to New Zealand, rabbits to Australia, eucalyptus trees from Australia to California, and opossums from Tennessee to California. Nutria and python snakes were accidentally introduced to Florida swamps. A few examples of unwanted species that have migrated are fresh water clams

Pollution

clogging pipes and water ways, moths infecting citrus groves, and Asian flu viruses.

Remarks

The environmental movement is to be commended for its effective actions reducing pollution and waste and for combatting the mismanagement of resources and the wanton destruction of land, forest, and sea. With no specified quality end point, however, abatement regulations continue past the point of stopping harmful practices to restrictions of little added benefit. Ecologists' goal of restoring wild life as it was in the past is only possible if modern society is barred from land where wild animals roam. The U.S. and world human populations have increased many folds, and civilization has spread to previously unspoiled areas. Predators and wild herd animals are incompatible with civilization, so humans and wild beasts can both survive only if kept apart. Ecologists fail to give adequate consideration to the economic and social consequences of their demands to return environments as they once were.

A part of all of us today would like to return to the scenic natural world in California that was described by John Muir and photographed by Ansel Adams. But the world they saw had no people obstructing their views. Today many thousands of visitors from around the world cram into Yosemite National Park, invited by the California

Tourist Board. They crowd to see the beautiful water falls from the valley floor and climb all over the majestic peaks. Automobiles create traffic jams and pollution in the valley and in the highways leading there.

In Muir's time there were thousands of people in California and a few hardy tourists; today nearly forty million people reside in California and millions more visit every year. The arid western lands cannot support millions of people and maintain the scenic wonders that once were. We cannot reverse technology nor, apparently, reverse the burgeoning world population. So we must learn to minimize and manage their destructive aspects while enjoying both nature and technology's benefits.

Due to general lack of understanding of science and how technology works, cures advocated by environmentalists often are not a satisfactory solution to the problems they have exposed. Often they advocate discarding a technology having an undesirable feature in favor of a new, unproven technology instead of correcting the original deficiency. All technologies have undesirable features which must be minimized in proper use.

Ecologists also seem to consider man-made chemicals to be artificial and especially toxic creations to be eliminated, and fail to recognize that toxicity is also present in many plant and animal species. Toxicity varies with concentration and the method of human contact. Many natural oc-

Pollution

curring chemicals, such as plant and animal toxins, lead, mercury, iodine, and arsenic are as toxic as synthetic chemicals. Many dangerous materials, whether poisonous or explosive, can be very useful, but must be handled with appropriate care.

Increasing quantities of consumer and industrial waste that is not biodegradable make burial in landfills undesirable. The human habit of trashing unwanted and waste materials is no longer acceptable for materials that do not degrade and are often hazardous. Recycling products whenever possible to extend their useful life and converting waste to innocuous materials are increasingly important.

9. Conservation

Many wars, including World War II, have been fought over the years to obtain critical materials. The technological developments of the 20th century have created a society that consumes enormous quantities of natural resources. As populations grow and the technology spreads to more nations, the demand for resources will increase further. So careful conservation of all these critical resources, and particularly forests and hydrocarbon deposits, is essential to maintain continuance of our global, high tech society

Arable Land & Water

Scientists say all of the arable land on earth now is used and less than three percent of the water is fresh. Growing populations encroach on lands and forests that are the habitats of wild animals or lands not suited for agriculture. Today nearly seventy million people live in the arid land west of the Rocky Mountains, over half in California alone. Expansion of populations into arid but fertile lands requires prioritization and careful management of water and land. Ownership, management, and control of water resources have been contested in the West since settlers first arrived. Supply of water has now become critical as large cities have grown up in arid areas of California, Nevada, Utah, and Arizona. Managing water resources and conserving arable land are of critical importance with the burgeoning population. Instead of the present piecemeal approach of reacting to each succeeding crisis and to the demands of special interest groups, California and the Western States need a comprehensive water management approach to balance needs of farms, cities, and ecology, while preventing salt water incursion into the aquifer along the shore lines. The old water rights laws now in place are not adequate to deal effectively with today's large population and complex water issues that affect many western states.

Environmentalists' attempts to restore nature as it was before people intruded are not realistic; burgeoning populations make this impossible. Animals must be protected in parks and reserves and forests managed for economic and social benefit to avoid wasteful practices.

Energy & Materials

Modern society is characterized by enormous consumption of energy created by the combustion of fuels. Reserves of coal, gas, and petroleum liquids are very large, but still finite. Increasing use of electricity generated by renewable processes reduces the need for fossil fuels, but electricity is not a total replacement. Oil and gas are necessary, not only to power aircraft and motor vehicles, but also as sources of carbon for producing chemicals, plastics, drugs, and materials that are essential for modern society.

Technological innovations have created need for many scarce materials, including precious metals, rare earth metals, uranium, lithium, precious gems, minerals and many specialty metals. While some critical minerals are spread widely over the earth, others are distributed sparsely. Possession and control of essential materials have always been a concern of nations.

Forests

Humans have been very destructive of forests throughout history. Trees have long been chopped down to provide building materials and fuel. Tall straight trees have been culled from forests to provide masts for sailing ships and poles and beams for buildings. Farmers clear cut forests

Conservation

to create arable land. Forests have been stripped of many hardwood trees over the centuries, primarily, to make beautiful furniture, buildings, and ships. Civilization is expanding into wild jungles, forests, and grass lands destroying the habitat of grazing animals and predators.

In the 20th century, old growth forests in North America were attacked to provide construction lumber for burgeoning cities and pulp for the paper industry. These forests, mostly owned by the federal government, contained some of the biggest and oldest living things, and were the habitat for many species of animals and birds. The forests in the East contained a mixture of maple, walnut, hickory, and cherry trees that are prized for furniture. Unfortunately, few of the old growth trees remain in North America. The green movement arrived too late to save them. To halt further destruction, environmentalists contested the destructive practice of clear-cutting which destroyed large swaths of trees and the creatures that lived in them. Such logging practices now have been largely curtailed, and the forests are being managed for constructive uses. The pulp and paper industries, major users of timber lands, now replace trees as they are harvested and recycle waste. Forests are important resources that must be preserved and managed for maximum benefit to society.

Oceans

High demand for seafood worldwide is leading to overfishing and decline in the populations of some popular species. Shellfish populations along ocean shores are particularly susceptible to overfishing and damage by pollutants. Some nations set strict harvest limits to protect supplies along their coasts, but no one controls the open seas. Conservation of the oceans and protection of sea life is a growing problem as technology's impact expands.

Oceans, that were considered endless, are being overfished and overwhelmed by pollution from thousands of huge ships moving freight and people. Commercial development of under water resources and navigation of submersible vessels is a growing hazard to sea life. Protection of the oceans is a growing international issue.

10. Global Warming

The Industrial Revolution resulted from the discovery in the 18th century that coal could be burned to produce energy, a chemical reaction. Pollution is a byproduct of coal from mining, handling, and from combustion. For many years, pollution from use of coal was treated as an irritant; the benefits outweighed pollution concerns. With expanding use of coal for electric power generation, the pollution could no longer be ignored. Pollution from coal operations became a major target of environmentalists

As natural gas became available in great quantities during the 20th century, it replaced coal for electrical power generation in many new plants. The main effluents from the use of gas for generation of electric power are innocuous CO_2 and water; there is little noxious pollution. When the Environmental Protection Agency was established in 1970, the auto industry was already working to eliminate all pollutants from the combustion process save for CO_2 and water (pages 91-92). Soon auto emissions were brought below the EPA limits. Preventing harmful air no longer would require eliminating fossil fuels. This upset environmentalists' justification for replacing fossil fuels with clean energy.

The theory that a planet's atmosphere could warm the planet's surface, the greenhouse effect, had been proposed in the 19th century. In 1957 Roger Revelle, Director of the Scripps Oceanographic Institute in La Jolla, California, offered the theory that CO_2 from burning fossil fuels might increase the greenhouse effect. Revelle moved to Harvard in 1963 where he established a Center for Population Studies. It was there that Al Gore was said to have learned of Revelle's global warming theory and the need for curtailing use of fossil fuels.

George Woodwell, a scientist at Brookhaven National Laboratory, became a board member of the Natural Resources Defense Council in 1970 and helped shape the NRDC's legal approach to curbing pollution and desecration of the environment. He was one of the first scientists to warn of fossil fuel emissions causing global warming. In 1979 he warned Congress that "we can't wait to prove that the climate is warming before stopping carbon dioxide buildup." [10] Woodwell felt use of fossil fuels upset the natural CO_2 cycle between plants and animals, and was a strong advocate for changing the process for creating energy.

About this time, Maurice Stone, a bureaucrat in the United Nations, became interested in the global warming issue and organized the first U.N. meeting on the environment in Stockholm in 1972. He headed the United Nations Environment Program (UNEP) in 1972 and was appointed Secretary General of the U.N. Earth Conference in Rio de Janeiro in 1992. Stone was instrumental in forming the United Nations Intergovernmental Panel on Climate Change (UN IPCC) in 1988 to produce reports supporting the global warming initiative.

Since pollution by motor vehicles had been greatly reduced by the actions of the auto manufacturers and fuel suppliers in the 1970s and 1980s, NRDC decided the best way to justify replacing fossil fuels was to have CO_2 declared a pollutant under the Clean Air Act. The logic was that although CO_2 was not harmful to people, increasing amounts in the atmosphere was harmful to the environment. The idea that CO_2 might cause climate change became a tactic to justify developing solar and wind power generation of electricity.

Environmentalists lobbied Congress for the next forty years to amend the laws without success during a period of mostly Republican administrations. In 1992 Al Gore promoted Revelle's theory in his book "Earth in the Balance".[1] Gore has since written several books arguing for restriction on carbon dioxide emissions to save the planet.

It is asserted that an increase in Earth's average temperature will cause the oceans to rise flooding coastal cities and cause the climate to be more changeable and damaging. The fact that carbon dioxide is naturally present in air in very small amounts, less than 0.1%, and is essential for the existence of flora and fauna, is ignored. Not surprisingly, the contribution of water vapor, the major greenhouse gas and present in the greatest amount, also is ignored.

The movement acquired renewed vigor in 2009, the first year of Barack Obama's Presidency, when the Democrats gained control of Congress. With few if any Republican Votes, Congress passed a law empowering the Environmental Protection Agency to declare that greenhouse gases, including carbon dioxide, are health threats and that the threat is caused by motor vehicles. The Supreme Court upheld the law in 2014 in a split decision. Governments and advocates around the world now have accepted Revelle's theory as truth.

Environmentalists profess that only the replacement of fossil fuels by energy produced from renewable natural sources-- solar radiation, wind, and biofuels-- will save the health of the planet. Electric automobiles must be developed to replace gasoline powered vehicles to eliminate carbon dioxide emissions. Government has responded by imposing regulations curtailing emission of carbon dioxide from power plants, and by providing incentives and subsidies to encourage development of electric cars. Governments are spending huge amounts of money, directly, as subsidies and loans to manufacturers to establish renewable energy processes, as well as imposing regulations that create higher costs on industry and consumers to hinder use of fossil fuels.

When it seemed Earth temperatures might not be increasing, the program was renamed Climate Change. No one can doubt that the climate is changeable.

The U.S. government provided tax incentives for development of biofuels to replace fossil fuels. Apparently, overlooked in the enthusiasm to eliminate fossil fuels is that biofuels, such as alcohol, produce as much carbon dioxide as fossil fuels and are less energy efficient. Environmentalists seem to ignore other sources of carbon dioxide, such as bio fuels, wild fires, volcanic eruptions, mammal emanations, and the constant wars.

Remarks

Environmentalists claim most scientists believe combustion of fossil fuels causes global warming by releasing carbon dioxide to the air, and those who do not believe so are deniers. Convincing proof has yet to be presented, however, that shows that the minute amounts of carbon dioxide in air cause global warming. There are better theories for Earth temperature changes over time than the release of gases from combustion of fossil fuels. Orbital, gravitational, and solar events are more likely causes of Earth's climate variations.

Geology, archelogy, and historical records indicate that Earth has long gone through extended periods of cooling and warming and extreme weather events before the human use of fossil fuels. Humans, plants, and animals have adapted successfully to many climatic conditions over the centuries. There have also been many periods of flood and drought and seismic disaster, but humans have endured and adapted to the changes. Climatic conditions vary enormously around the earth at any time. Temperature and climate variations that are often violent are usual occurrences for the planet. Cooling and warming periods have extended for periods of hundreds and thousands of years.

The uncertainties establishing cause/effect relations between correlated variables are discussed in chapter 12. Do minute changes in average Earth temperature really affect weather or does weather affect Earth temperature? Is there an affect at all? Can carbon dioxide change the weather? Science knows much more about the universe than about the Earth. Recent studies of the atmosphere are just a beginning of meteorology as a science. If humans have an effect on the weather, it is more likely to come from the destruction of forests and the creation of megacities paved with concrete. The years of expanding use of fossil fuels have seen a large population shift in the

U.S. from the urban East to a suburban West creating substantial changes to the environment.

"Climate Change" is not likely such an urgent threat to humankind as to warrant curtailment of use of fossil fuels. Development of replacement for fossil fuels by harnessing solar energy is still in its infancy and will take many years before it is economically competitive. The intensity of solar energy striking Earth is low (fortunately for living things); so it is unlikely that enough energy could be harvested to meet the enormous needs of society. In addition, the economic and social problems connected with moving from petroleum to a solar power based social structure would be enormous. Achieving control of global warming, even if possible, could not be attainable for many decades.

The tragedy of demonizing carbon dioxide emissions is that it has diverted efforts from ecology and conservation approaches that could provide more immediate benefits. It has also saddled the economy with unnecessary costs. The alcohol addition to gasoline has increased costs for the motoring public and disrupted the economy of the Midwest farming states. Money spent to create an electric "fueling" network along highways and within cities might be better spent on improving highways and vehicle energy efficiency.

Technical Issues

Most major technologies, such as harnessing the power of the sun, can be very complicated and depend on the successful development of supporting technologies. Commercial utilization of solar involves the separate issues of collecting solar energy, which is spread thinly and unevenly across the Earth surface, storing the energy until needed, and distributing the energy to the ultimate power users.

Little if any thought was given to the complexity of effective utilization of solar power when government incentives were initiated in the 1970s. The radiant heat of the sun was collected for water heaters and swimming pools. Suitable durable materials for absorbing sun's heat could not be developed. Venders went out of business, government incentives were lost, and home owners, who scrapped the equipment installed on their roof tops bore the costs of failure.

Utilization of the sun's energy got a big boost by invention of solar cells that convert sun's radiance directly to electricity. The cost-effectiveness of these electronic devices has increased greatly in the last few decades, and use is spreading. The long term durability and performance in diverse weather conditions remains an issue. Solar cells are uniquely attractive for creating electricity near the point of use. A great disadvantage is that the cells work only when the sun shines. The economics of solar electricity production is difficult to assess because of government subsidies and regulations directed to force the use solar power in the electric grid. Electricity is a very useful form of energy, but it has critical limitations for use, particularly, in transportation.

Political Issues

Government advocates around the world have accepted Revelle's theory as truth. Many scientists, however, consider information from solar, geological, and astrological observations more convincing than the theory that human activity causes global warming. The Heartland Institute in Chicago is a longtime critic of climate change and sponsor of dissenting publications and scientific conferences.

Al Gore has explained that many people resist the "Inconvenient Truth" about Global Warming because we are going to have to change the way we live. "Only meaningful and effective solutions to the climate crisis involve massive changes in human behavior and thinking." [11] This certainly would be the case. But such sacrifice is completely unnecessary if the environmental focus would return to eliminating pollution and waste and stop trying to force technology (and society) towards an uncertain future.

President Obama recently lamented that those opposed to global warming seem to be all Republicans. This is not surprising since this is a political debate involving misuse of technology and not a scientific issue. Global warming has developed the religious fever of believers and non-believers.

The global warming movement now is threatening dictatorial actions to force acceptance. Recently, it was reported that "California Attorney General Kamala Harris is investigating whether ExxonMobil Corp repeatedly lied to the public and its shareholders (in the 1980s) about the risk to its business from climate change — and whether such actions could amount to securities fraud and violations of environmental laws (Ventura Star, Jan.21, 2016)." Then in March, 2016, at a Senate Judiciary Committee hearing on Justice Department operations, Attorney General Loretta Lynch acknowledged that there have been discussions about possibly pursuing civil action against climate change deniers and that the matter had been referred to the FBI for investigation. Our nation is entering a dangerous path if dissenters are penalized for their views. It appears Global Warming is taking on a modern version of the fervor of the Spanish Inquisition.

Sadly, no one seems to have listened to President Eisenhower's warning (page 60) of the danger from the formation of a powerful network of government funded experts and a scientific-technological elite, assuming undue influence. We are paying a huge price for ignoring this advice,

Return to Nature

Underlying the movement to eliminate the use of fossil fuels is the dream to return to nature and eliminate artificially produced materials. We all enjoy visiting nature, but few would sacrifice all of the benefits technology provides to return to live in the world preceding the 20th century. We cherish the ability to fly and drive to those pristine nature reserves, stay in comfortable hotels, and ride lifts and gondolas to take us into the back- country.

With modern technology, renewable energy sources have many economical uses, but electricity is not a replacement for fossil fuels. It is not adequate for most motor vehicles. Science and innovation do not follow dreams very well nor authoritative direction. It is ironic that those wishing to return to nature have embarked on a program to control the weather.

Part IV

Human Nature, Science, and Technology

11. Human Nature
The Innate Forces That Drive Us

Science has learned much about the workings of the natural world, the evolution of civilization, and human social behaviors. As we consider the effects science and technology have on social institutions, it is instructive to consider what makes humans think and act as they do.

In our minds the natural world is beautiful and idyllic, and we long to return to this unspoiled life. In truth, however, the natural world is violent and changeable. Cosmic rays, space dust, and radiation from the sun, constantly, pelt the earth. Meteorites burn up in the atmosphere with great brilliance, and on occasion a large meteor strikes the earth with explosive force. Earth is shaken frequently by destructive earthquakes as the tectonic plates grind against one another. Volcanos spew noxious fumes and ash into the atmosphere and exude molten rock across the landscape. Volcanic mountains rise from the sea while others are blown apart. Occasional storms rage across the world causing localized floods and landslides, while gentle rains bring forth beneficial plant life. Periodic droughts and plagues have caused havoc to local populations as far back as is known.

Animal and plant species continuously evolve as they adapt to their changing environments. Insects and microbes have very short lifespans and mutate rapidly. Large plants and animals have long lifespans and mutate more slowly. Population growth of species is limited by consumption of necessary resources, by reaching impregnable borders, or by destructive plagues. Many species viciously prey on those below them in the food chain and fight for dominance with other predators. Although Earth is a chaotic place, it is the only known spot in our solar system, and maybe in our galaxy and beyond with the "Goldie Locks" celestial conditions needed to support life. In the last forty thousand years or so humans have grown to dominate the animal kingdom.

Human Behavior

Humans share many traits with animals; indeed, human DNA differs from that of mammals by comparatively small amounts. Humans also share individual and social behavior characteristics with some animal species. Humans are distinguished from animals by two remarkable features: enormous brains and modulated speech. The brains make humans rational, inquisitive, and inventive. Modulated speech fosters social structures and the ability

Human Nature

to transfer learning to others and to subsequent generations.

Early humans depended on their environments, and used what was available for food and shelter. Materials were limited to those derived from plants, animals, and the Earth itself. They grouped together in family tribes for protection from predators and other humans. Energy was limited to the strength of people and their animals. Communication was verbal and spread by messengers. They could travel as fast and as far as they could walk or float on streams and lakes.

Primitive societies throughout the world have been structured as tightknit, extended families. Small families combined to form tribes and clans, usually led by a dominate male leader. Males protected the clan from competing clans and sired offspring (often polygamous). Males inherently dominated, ambitiously conquering obstacles, and often competing viciously for power. Females gathered together to nurture and protect the offspring and provide sustenance. Instinctively, humans are vicious bullies striving for dominance over their peers; they are gregarious and territorial and can be brutal predators. They are curious, "going where no man has gone before." While fearful of the unknown, they take chances when risks of failure are understood.

Unique cultures evolved in isolation separated by geography, but shared similar social structures. Clans were fiercely territorial and possessive. Speech variations and beliefs distinguished each clan from others. While initially they were hunter-gatherers. They evolved to agrarian societies living in fortified cities - they became civilized. Conflicts occurred when societies collided, culminating in devastation, massacre or enslavement, and loss of wealth of the loser. Encroaching civilizations conquered and forced out weaker often nomadic tribes.

We all possess such innate qualities as curiosity, inventiveness, competitiveness, possessiveness, and gregariousness. We are fiercely territorial. Humans also seem to have a "sheep" gene, deferring to people they believe wiser or stronger. They obey and follow leaders, sometimes naively, because they want to belong or must obey the leader's message. These qualities may be more dominant in some individuals than others (women are usually more social and caring than men). Civilizations create rules for humans to live together peacefully and authority to enforce the rules; this never seems to succeed fully in taming the natural belligerent instincts in humans.

Civilization advanced by utilizing the materials that environment offered to improve living conditions. Curiously, throughout history humans have valued highly gold and gem stones for decoration even though they have no practical utility. Viking raiders plundered European castles and towns for their valuables. Alchemists strived to convert base metals to gold. Spanish conquistadors plundered the native civilizations of the Americas for their gold. Adventurers rush to discoveries of new gold deposits to claim a share. Why do humans prize trinkets so highly?

We differ individually and socially by our experiences, adopted values, and beliefs. Values and beliefs can be irrational in the sense that once accepted, they are very difficult to change by argument. We defend our beliefs strongly and try to impose them on others. We need "values" to provide direction to our lives. Innate behaviors may be suppressed by values and beliefs, but often surface as socially unacceptable behavior.

Humans are highly competitive, individually and collectively. We strive to be the best; we want to own the most, the biggest, or the most unique. Winners and champions are celebrated. We constantly compare ourselves to others, "to keep up with the Jones". Cities strive to build the tallest buildings and the biggest arenas. Companies and countries compete to build the biggest ocean liners and cargo ships, the biggest and fastest airplanes, and the swiftest trains. We maintain extensive statistical records of sporting contests, striving for new record performances. Daily weather data are tracked for micro climes and compared going back as far as records allow (seldom more than one hundred years). Economic statistics are compiled for federal and state governments to measure the economic health of the country.

Human Experience

Humans are born alike with a remarkable brain that is programed for great things but is initially void of data. Our lifetime of experience provides the information that directs our functions and makes us unique. We each view the world with eyes that see what our experience directs them to see. Analogous to computer jargon, we are born with brains having an "operating system" that is receptive to many "application programs" which define our diverse cultures. We are biased, and have difficulties understanding people of different cultures and experiences. We are taught facts and opinions, but learning occurs only with experience and the acceptance and absorption of beliefs.

We are the products of our individual experiences. To a businessman or banker, every problem can be solved with money; to a lawyer problems are solved by a law to change the rules; a politician takes a poll to see what the electorate wants and then votes to please them.

We frequently encounter people here and abroad from diverse cultures speaking many different languages. We are also faced with the special idioms and jargons of technologies, businesses, finance, and social groups. Words often have different meaning for speaker and listener because they have different experiences. Generally, people talk and opine but do not listen well; yet listening is the basis of learning.

Effective communication is more difficult today than ever. We admire orators and "communicators", and we hear, but often, do not listen well. Today technology has overcome the geographical barriers that separate societies; so that clashes of people and ideas are now worldwide.

Organizational Behavior

Humans and animals associate in groups for security and to do things that cannot be done individually. It is the nature of humans with common goals and interests to assemble in groups and follow the dictates of strong leaders. Personal interests may be suppressed in favor of the group's goals, but extensive diversion of personal interests weakens the group's effectiveness. Organizations behave much like individuals and possess innate and learned qualities and purposes just as individuals. Organizations, once created, are living creatures whose principal purpose is to perpetuate them-selves. It is very difficult to disband an organization once created, particularly large organizations, because of high economic and personal costs when the group is disbanded.

Mobs are short lived, unstructured organizations in which individuals feeling anonymous behave in ways they would never do alone. Innate behavior dominates.

Understanding

Language

Communication requires mutual understanding between speaker and listener who often have different experiences and thought processes. Language composed of words is the process of communication; speaker and listener must have common understanding of the meaning of words to convey information accurately. Languages are very dynamic, changing as societies change. Words are added and fade away over time. Many words have several meanings and meanings may change over time. Some concepts are difficult to translate from one language to another. The meaning of nouns and verbs is usually well understood, but adjectives and adverbs which qualify thoughts and concepts can be confusing. An example is *average* and *normal* that is often used as synonyms in describing the weather and other phenomena, but have quite different meanings. Strictly speaking, *average* is a mathematical calculation and may not actually exist. *Normal*, however, is the most prevalent number or range of numbers. Usual or most common can be significant; but average explains little. Percentages, widely used in reporting, also are very misleading unless qualified by the numbers they represent (an insignificantly small number increased by 50% is still insignificant, but the national debt increased by 50% is alarming).

Communication
Speakers often don't realize the impact of their words. What listeners hear may not be what speakers intend. If the listener doesn't seem to understand, the speaker often repeats his message louder and then may force his views on the listener (examples are parent to child and political views). In group discussions speakers sometimes talk simultaneously; so nobody is really listening (notably TV talk shows and panel discussions). We tend to talk and listen to those with similar views and tune out those with opposing ideas, particularly in politics.

Opposing beliefs are major obstacles to cooperation. Understanding others has always been difficult, but technology has created a new world of specialists having greatly differing training and experiences that compound the difficulty to understand. We encounter many people here and abroad from diverse racial and religious cultures speaking many different languages. We are also faced with the specialty languages of professional disciplines, such as chemistry, electronics, accounting, medicine, and finance, as well as jargons and idioms of politics, music, and sports. Effective communication is more difficult today than ever.

Fear and closely held beliefs are major obstacles to communication and understanding. In earlier times people looked to their gods for explanation and protection from unexplainable things. People deify or demonize what they do not understand. Chemicals, which are considered artificial, are bad, but in today's complex society scientists, whose activities few understand, are held in high esteem, [5] and their views are accepted as fact. Scientists, seldom having competence beyond their specialty, are no better at foretelling the future than the sages of the past.

Humans shut out the noise that bombards their senses and focus attention narrowly. This is essential to have a conversation in a noisy crowd or factory. But narrow views restrict analysis and impede communication. This trait combined with the innate need for action leads to poor decisions. To a deer hunter all that moves in the forest is a deer, sometimes to the detriment of fellow hunters. The cure can be worse than the ailment. Doctors used to bleed the sick to remove bad blood. Environmentalists say carbon dioxide released by combustion causes climate change.

People look to authorities, whom they trust, for guidance: parents, clergy, teachers, government, or any orator with a message that reflects their values. When advice is proven wrong, respect for authority is lost, and bad things happen. Many institutions have lost credibility with the people who depend on them.

Wisdom

Wisdom and good judgement seem not to be genetic but must be learned by experience. Knowledge does not bring wisdom; wisdom is using knowledge for betterment and for recognizing what is betterment. Science and technology have created great things, but people seem no smarter nor inventive than those who have gone before. People certainly have much more information available at the touch of a button or a query to "Siri" or "Alexa" than those less connected, but we seem none too bright in employing the knowledge to improve our lives. Cultural, religious, and personal conflicts are common. We abound in theories about the causes of natural events and paths to greater health and longevity, but facts and proof are elusive. After sending a man to the moon, we think technology can do anything if we wish: split the atom, send space ships to other planets and galaxies, and even control of the weather. We probably can't do everything we think we can, nor will many new things make our lives better. Wisdom is the understanding of choices and selecting the most beneficial to ourselves and our posterity.

Written Language & Symbols

In addition to many cultures and languages that divide peoples, there are written languages and symbols that further complicate communication. Ancient civilizations separated by geological barriers, developed unique languages and written records that varied from pictorial symbols in Egyptian hieroglyphics and Chinese pictographs to alphabets of symbols representing spoken sounds in Greek and many modern European languages.

Ancient philosophers introduced symbols for numbers for counting and for algebra and geometry derivations. Later Arabic symbols for zero to nine created the base ten numbering system that enhanced mathematical calculations. A binomial system was developed in the last century to operate the electronic machines in the present digital age. Several hundred years ago, musicians created a graphic system to record compositions for later performances. Development of atomic and molecular theory led to symbols for the elements and classification in the atomic chart in the 19th century. In the 20th century, the new science of chemistry created a graphic symbolism to describe the molecular composition of matter. Physicians developed models of the human body to illustrate the locations and features of all the organs.

In all cases these special written "languages" aid specialists' understanding and performance but increase the difficulty for the untrained to comprehend. Consequently, specialists, often, are held in high esteem by those who are not familiar with their specialty. In the present specialization society, common understanding is a rare commodity. The significance of specialists' opinions is often exaggerated.

The Elusive Middle Ground

Humans seem to have inherent social behaviors that impede finding a middle ground on issues that competing factions can support. Every issue is viewed from the extremes: right or wrong, win or lose yes or no, left or right. Pursuit of the extremes leads to results that only one side can condone.

Today, Democrats are striving for a government which is paramount in society, taking money from the producers of goods and services and distributing the proceeds among the people. Republicans, on the other hand, strive for a much smaller government that allows the producers of goods and services to distribute wealth to the people directly with minimal governmental regulations. The first dictates what people need but stymies the innovation which leads to economic growth; whereas, the second stimulates innovation and generates economic growth, but provides less social support for the unfortunate. The first leads to unsustainable deficits and constrains individual freedoms; the second leads to more stable government and greater personal freedoms and responsibility but a reduced social safety net. We need to find a middle ground that distributes benefits fairly among the populace while maintaining free enterprise for economic growth. Extremes are never good.

12. Science & Technology

We are proud of the great technical accomplishments our country has achieved since its founding. We treasure the inventions of the many known and unknown contributors who created our present way of life, and expect even greater future advancements. Few people, however, understand how science and technology work and how they came to be. This confusion and misunderstanding arise from the complexities of our specialized society and the unique nature of the scientific and technological processes. Science and technology are viewed as kindred functions, whereas in fact, they differ greatly in operation and purpose.

A Brief History

From the beginnings of civilization, humans have tried to understand the dynamics of their environment. The ancients were fascinated by the stars, moon, sun, and the earth around them and speculated about how things worked. Early societies created religions to explain human existence and the workings of their environment and to set moral codes. Philosophy, art, drama, and the begin-

Science & Technology

ning of scientific thought evolved and were recorded for posterity. There followed the beginning of Science - the philosophy searching for rational answers to nature and natural phenomena. When they observed the spatial orientations of the heavenly bodies and measured their movements, astronomy (the first science) was born.

Modern science had its beginnings with the Renaissance in the 17th century when Galileo applied mathematics to physical phenomena and tested theories by experimentation. Newton followed with his laws of motion. After two hundred years of thought and experimentation, huge strides were made in the 20th century with understanding of the molecular structure of matter and how bodies interact - the creation of Physics and Chemistry. This was achieved largely through the efforts of such European theorists as Max Planck, Niels Bohr, Werner Heisenberg, and Albert Einstein. Their theories of Relativity and Quantum Mechanics explain the operations of the universe from atoms to galaxies replacing Newton's laws of motion. The physical world is now well understood, although experimentation continues to explore the depths of the universe and the internal composition of atoms. Astronomers believe they have found the ends of the universe and how it began - the Big Bang theory.

The 19th century was a period of discovery in the biological sciences and medicine. Darwin's theory of evolution explained the variation in plant and animal species. Hybrid plants were created. Germs were shown to be a cause of illnesses and spoilage of foods. Great strides were made in the 20th century to improve health. Sterilization and sanitation practices were established. Medicines to combat devastating plagues and extend life were developed. Anesthetics and analgesics were created to reduce pain. By applying the principles of physics and chemistry to biology, the molecular structure of DNA was determined and shown to be the factor that differentiates all organisms and directs the genetic reproduction process of living species. Much remains to be learned, however, about the complex chemical, physical, and electrical processes that make living bodies function.

The petrochemical industry that brought on the Carbon Age was created in the U. S. in the years following World War I. During the Second World War, the demand for synthetic materials to replace scarce products like natural rubber spurred the industry's growth. Following World War II, a great expansion of the industry took place over about thirty years. Many companies in health care, communication, energy, plastics, textile fibers, etc. built large laboratories for research and for development of products and processes. Some products replaced natural materials providing better performance at lower cost, while many more were unique materials that made possible the rapid expansion of electronic and space technologies so evident today.

Material sciences evolved to make new structural modifications of metals, ceramics, semiconductor elements, and composites. These new materials together with those derived from petroleum provide the basis for all the modern technological developments. Today we see the products of technology (planes, automobiles, computers, satellites, movies, and medical miracles), but few realize these could not exist without the materials developed through chemistry and physics. Modern society cannot exist without products from coal, petroleum, and natural gas.

Science

The Oxford Current English Dictionary defines science as "the systematic study of the structure and behavior of the physical and natural world through observation and experiment". Science is the branch of philosophy seeking knowledge of nature. The foundations of science are the two disciplines: physics, concerned with the properties of matter and energy - "How" the universe functions; and chemistry, concerned with the composition of substances and how they interact - "What" matter is. Science does not try to answer "Why" which implies intention; Why is the provenance of Religion.

Science progresses in several phases: exploration first, followed by classification, measurement, and testing of theories. Science seeks truth and understanding. Science is never ending - there are always more questions to answer.

Universities and technical institutes traditionally focused on basic or theoretical research to learn how things work, whereas industrial laboratories stressed exploratory or applied research seeking materials and processes of economic value.

Prior to World War II, the U.S. government was not involved in scientific research. Industrial laboratories formed in the early part of the 20th century conducted large research programs supporting their businesses and funded fellowships at many universities and technical institutes. Professors who were retained as consultants provided an interface between education in basic science and the commercial world.

During World War II, commercial and academic research institutions provided the scientific and engineering resources for the many military programs including development of radar, penicillin, and the atomic bomb. The Office of Scientific Research and Development headed by Vannevar Bush, was created to coordinate the many programs. The warring nations focused on developing weapons and technical advantages to triumph over their adversaries. Arms races ensued to develop more lethal weapons, bigger and faster warships and airplanes, and control of critical natural resources such as oil and rubber. Arms races reached a climax after defeat of the Axis Forces in World War II as the U.S. faced off against the

Science & Technology

U.S.S.R. over control of atomic weapons and utilization of space.

Concerned that the government had to draw people from the private sector to manage scientific projects for the military during the War, Vannevar Bush recommended that the Federal Government sponsor research in the physical and medical sciences that is important to national defense. Congress, subsequently, created the National Science Foundation in 1950 followed by the National Aeronautics and Space Administration (NASA) in 1958. While the military developed ballistic missiles and satellites in secret, NASA addressed peaceful and public uses for space including communication satellites and, space probes. The U.S. government continues to provide considerable funding to universities and research institutes mainly for space exploration, nuclear physics, medical research, and recently atmospheric research. Research is still conducted by large pharmaceutical, chemical, and electronics corporations.

Technology

Technology is a term that is widely used today, but with several different meanings. In the investment industry technology, refers to those companies, products, and systems in the electronic and cyber world. The word technology also is used often to refer to specific processes and products.

The Oxford Current English Dictionary defines technology as "the application of scientific knowledge for practical purposes". This restriction to scientific knowledge is misleading. Science had nothing to do with the invention of such things as the cotton gin, automobiles, telegraph, and refrigeration. Ancient civilizations built massive structures, cities, and fleets of ships centuries before science made its appearance. Even energy was created before it was understood that a chemical reaction was the cause. An exception is the electronics industry which evolved from the study of semiconductors that led to the invention of transistors. The popular idea that science leads technology, generally, is incorrect.

A better definition would be "technology is the application of *knowledge* for practical use and economic gain". Technology is not dependent on science. Technology is a kit of tools, analogous to mathematics. Its value depends on how it is used. An engine creates more economic value when it powers a factory producing goods, than when it powers a race car achieving performance records. Technology makes things cheaper and easier requiring less labor, or it creates new desirable things. Technology is more akin to commerce than to science. Engineering is the discipline that develops practical uses of technology. Technology may be spectacular, but provide little economic benefit – change and difference are not always better nor more valuable.

Science & Technology

Development of major technologies can be a very long and complex process involving many inventors, machines and products. To be successful major technologies must be timely-- all essential materials and subordinate technologies must be available to supply a receptive market. The story of the development of steam engines leading later to internal combustion engines summarized in Chapter 1 illustrates the complexity and spontaneity of the technology process. It involved new materials of construction, new fuels, many engine designs, many applications and the contributions of hundreds inventors and engineers. It began with a machine to pump water and ended with the family car that led to today's suburban living. It led also to engines that powered aircraft to supersonic flight, and to today's air transport industry. The development of the machines that power modern transportation was a process that took more than two centuries from the time Thomas Newcomen made the first steam engine.

People attribute the amazing technical achievements of the 20th century to science, and believe scientists can accomplish anything they set out to do. However, the great aerospace achievements arose by pushing the envelope of present knowledge, not by new scientific discoveries. They were achieved by developing greater power, better materials, and improved control systems - by engineering not scientific discoveries. Engineering put men and materials in space; scientific knowledge directed the rockets to the proper orbits and space landings. The present high tech industries have developed from application of scientific discoveries fifty years ago.

A major technology accomplishment in the twentieth century, that may not be generally recognized, was improvement in the quality, quantity, variety, and safety of food and water. Chlorination of the water supply nearly eliminated diseases such as typhoid fever and cholera. Improved sanitation, preservatives, and vitamins improved food quality reducing the occurrences of malnutrition and intestinal disorders. Fluoridation of drinking water has greatly reduced tooth decay. Refrigeration enhanced quality of food during shipping and storage. Hybridization increased the quality and variety of many fruits and vegetables, including the creation of Russet potatoes and Santa Clara plums. Cooking oils were introduced, reducing use of traditional lard and butter. Improvement in food and water quality is a major factor for increased life expectancy and the larger size of people today.

Food is now plentiful, affordable, tasty, and fattening. So people eat more even as they become more sedentary in their jobs. Obesity has become a national issue. Top chefs teach fine cooking and restaurants abound with low cost tasty food; yet nutritionists tell us to diet and eat healthy; fitness gurus tell us to diet and exercise; and even governments tell us what to eat. To counteract excesses from over eating of rich foods, people adopt fad diets and consume food supplements to correct presumed dietary deficiencies and increase life expectancies.

In the past, food technology was directed at eliminating acute illnesses caused by consumption of tainted food. Now food science is directed at preventing degenerative disease from consumption of ingredients that may adversely affect various body organs. Medical science is seeking the same result with drugs. Some people want to eat "organic"- avoiding foods with additives and artificial treatments, but this is where we began a hundred years ago with tainted food!

Technology Dark Side

Technologies may find useful applications beyond those that were intended; but technologies also often have unintended and unwanted consequences, a "dark side". It seems to be a natural law that something unwanted and unanticipated will accompany every new technology. Examples are the side effects of drugs and biocides, waste disposal issues, dangers of toxic exposures, and accidents from careless handling. Many technologies also have dangerous consequences when misused deliberately, such as weapons and internet scams. Successful technologies balance benefits and unwanted consequences. Sometimes society has discarded useful technologies because of undesirable side effects instead of using them in ways to avoid the unwanted feature.

Scientific Process

Research Methods

Ideally, the scientific process involves verifiable cause and effect experiments to prove theories. It is mathematical and analytical. This method was used to make the spectacular advances in physics and chemistry of the 20th century. Such cause & effect experiments, however, are difficult or impossible to perform in studies of biological processes, nutrition, and natural phenomena wherein measurement and control of variables are imperfect. Researchers must rely then on anecdotal and statistical studies. Statistical methods correlate selected variables and results. It should be understood, however, that correlation does not prove that one variable causes another or even whether there is a cause and effect relationship at all. Computers today make statistical correlations easy, maybe too easy, since conclusions are only as good as the logic in the software and the choice and quality of the input data.

For clinical drug studies the statistical process requires large populations of subjects to establish efficacy of cure of ailments, method of application, and dosage. The statistical method tacitly assumes a uniform population. Since humans vary by sex, age, physical condition, and genetics, subjects usually react differently to treatments so the composition of the studied population must be carefully controlled. Results are expressed by percentage of successful cures and come with warnings of observed side effects. While such research is costly and imprecise, many fine drugs have been developed to treat and cure acute diseases.

When it comes to prevention of chronic diseases or assessing the long term health effect of exposure to traces of toxic substances, scientific proof is more difficult to assess. There is wide reporting in the popular media of individual research studies aimed at prevention of illnesses and improvement of health and longevity. The significance of these studies for individuals often is qualified by such words as could, might, may, or should. Conclusions of early studies may be changed or disproven by further studies; so conclusions from initial experiments should be considered skeptically until corroborated by further investigations. The prevention of future harm or occurrence is not apparent for many years; so long term studies are required. The efficacy of drugs is tracked for years after initial introduction to confirm performance and to serve as a base for comparing treatments.

The scientific basis for nutrition is particularly weak. Recommendations for healthy diets and nutritional supplements are largely anecdotal. Cures for nutritional deficiencies can be monitored effectively, but health benefits of larger doses of supplements are largely speculative. Benefits for users vary widely. Although proof often is lacking, people adopt these recommendations with the attitude that they can't hurt and may help. People have been promoting tonics for better health and telling their children what is good for them to eat for centuries. The scientific basis for the present industries that have evolved to provide dietary supplements and pills for good health is not much better than that of the past. Scientists are beginning to address this human variability in nutrition studies.

Size, Place, & Time Matter

The characteristics of matter and chemical reactivity are affected by temperature, pressure, surfaces, volumes, concentrations, and foreign contaminants. Generally, chemical reactions are enhanced by increasing temperature and pressure. A chemical process designed to operate at sea level must be reengineered to operate in mountain cities where atmospheric pressures are lower. A laboratory process in glass vessels must be redesigned to operate in large production steel vessels having much smaller surface to volume ratios. Small concentrations of foreign materials may catalyze and enhance chemical reactions or, conversely, impede the onset of reactions.

Chemical structure, properties, and reactivity of molecules at surfaces and in very thin films differ from that in the interior of substances. Scientists are studying "Nano" films of one molecule thickness to understand how their properties differ from bulk properties. Some biological processes may function by "Nano" surface chemistry.

Science & Technology

Effectiveness of medicines depends on dosage: too little provides no benefit; too much is toxic.

Electricity is easy and safe to handle in the small amounts used for portable electronic devices. It is more difficult and hazardous, however, to store and use electricity in the large quantities needed to power heavy machines or to be shipped through the power grid.

The destructive effect of humans on the environment is minimal for individuals but enormous for crowds. The huge world population today creates issues that were unknown in past centuries when populations were much smaller.

Solar power is spread very thinly and unevenly across the surface of Earth. So it must be accumulated and concentrated to provide large quantities of electrical power.

Lifespans of microbes, insects, mammals, and plants vary enormously from minutes to hundreds of years. Similarly, geological and astrological times range from hundreds to billions of years. Studies must cover time periods within the ranges of the lifespans of subject matter to have meaning. Many global warming studies cover much less than one hundred years, too short a period to provide significant predictions of future planetary behavior.

Contaminants, Toxicity, & Risk

When Leeuwenhoek first looked through his microscope lens in the 17th century, he noted "wee beasties" swimming in water. He named them "animalcules"; they are now called microorganisms. Today we know that few things are truly pure, that is without contamination by germs, dirt, chemicals, liquids, or gases. When viewed microscopically, our world is not a very clean place.

One of the great achievements of science has been the ability to detect, characterize, and measure trace quantities of contaminants in materials, water, and air in parts per billion and even less. Such tiny amounts were not detectable just sixty years ago. People are alarmed now when traces of chemicals that are toxic in larger doses are detected in things they might ingest or handle. We are constantly exposed to minute amounts of toxic contaminants in the food we ingest, the water we drink and wash with, the air we breathe, and the cosmic rays that cascade constantly upon us. It is important to understand that toxicity of a material varies widely among animal species, insects, and plants, and by the kind of exposure: skin contact, injection, ingestion, or inhalation. All materials can be toxic in very large amounts. Some materials, like asbestos fibers and silica dust, are much more toxic when inhaled than ingested. No matter the method of exposure, a toxin must exceed a critical level to be harmful. Although toxicity is comparatively easy to establish in large doses, it is very difficult to measure the effect of long exposures to low doses. Studies are mostly anecdotal and require decades to follow long term health effects. It seems that human bodies are designed to function in an impure world. They adapt to exposure by light, heat, microorganisms, toxins, and cosmic radiation by internal heating and cooling mechanisms, skin composition, and immune systems.

It should be recognized that every activity has risks and dangers. Without risk there is no gain. Prudent practice is necessary, but risk cannot be totally eliminated without stagnation. People fear the unknown and what they don't understand, and modern science has created many things that are hard to understand. There is widespread fear of cancer, birth defects, and other degenerative ailments whose causes are unknown. Science has created many materials and chemicals, often having complex and strange names, that could be harmful if misused. Compounds of mercury, lead, and arsenic which occur naturally in the environment and have been used since earliest times for medical and industrial purposes are now considered hazardous even in trace amounts. Many devices are also in common use that may expose people to small amounts of radiation from micro waves, X-rays or radio waves. Generally, proper safe handling practices, effective dosages, and toxicity levels are established for these products before release to the marketplace. Small amounts of toxic materials should be treated prudently with respect but should not be feared.

Tools & Significant Numbers

Modern scientific research is aided by many instruments and processes that were not generally available 50 years ago. Computers have replaced slide rules for the manipulation and recording of experimental data and made books of tables of physical properties and mathematical calculations obsolete. Chromatography and spectrometers have reduced need for distillation and crystallization for separation, isolation, and purification of materials. Electron microscopes and nuclear magnetic resonance machines enhance the study of molecular structures.

There is a tendency for the media to report results of individual experiments published in scientific journals and to predict the future significance of the study. It should be recognized that an experiment is one data point, one fact. Facts are analogous to pixels; it takes many to complete a picture - to establish truth. Just as missing pixels create a distorted or misleading picture, missing facts distort the truth.

With instruments capable of detecting and measuring parts per billion and less and computers able to manipulate masses of data, results often are reported to unwarranted accuracies. Experimental results are no more accurate than the least precise measurement. Reported studies seldom indicate the precision of the results.

Science & Technology

Gene Manipulation

The discovery of the composition and function of DNA has brought science into the moral dilemma of the creation and manipulation of living things. It has been possible to create a living animal from DNA and to modify a specific characteristic for improved benefit. Hybridization of plants and breeding of animals are long accepted processes for modifying genes to improve domestic species. The process of recombinant DNA, that alters specific genes more precisely than hybridization, is considered unnatural and less healthy by some consumers. Nevertheless, genetically modified crops, now in widespread production, have become basic components in the food supply.

Stem cell research, whereby human cells are grown to create living tissue, violates the religious beliefs of some people. Scientific creation and manipulation of living things for greater performance and longer life remain a great moral challenge for mankind.

Innovation

Innovation is an inherited quality of humans. We all excel at solving problems within our education and experience capacity. History teaches that people with similar experience will approach a problem in the same way and hence arrive at similar conclusions. Thus innovation is fostered by teaming people of different experience to "think outside the box". Probably, the most innovative scientific research organization was Bell Labs, a division of AT&T established in 1925 with the mission "to connect everyone on earth." [12] Bell brought together top scientists and engineers from diverse fields and provided them free reign to carry on their own research related broadly to "communication". Their useful inventions were manufactured by the Western Electric Company, an AT&T subsidiary, and put into operations by the AT&T regional phone companies. The government had long contested ATT's monopoly over the telephone industry. So the company was finally fragmented, and Bell Labs was disbanded in 1970. At its peak in the 1960s, Bell Labs employment totaled 15,000 including 1,200 PhDs. Scientific innovation was greatest from the late 1930s to 1970.

Over the years of their existence, Bell Labs invented long distance and cellular telephones, transistors, silicon semiconductors, fiber optics, and "Information Theory" that led to today's electronic and computer industries. They also introduced microwave phone transmission, communication satellites, lasers, and solar power cells. Their patents were provided to other companies, license free. Texas Instruments, Intel, and other large electronic companies were founded on Bell patents.

Science & Technology

Many new products arise from wars, where ingenuity is focused on bettering the enemy and overcoming operational problems. The arms race during the wars of the 20th century in Europe, and the space race in the Cold War were major factors in the development of modern Technology. The adage "Necessity is the Mother of Invention" is true. Technology is innovation that works to solve practical problems.

Remarks

The popular confusion between science and technology derives from their use of similar methods. The difference between the two disciplines is the purpose or goal of the investigations. Science's objective is knowledge and technology's objective is useful products and processes.

Confusion comes when the media report scientific studies, often culled from scientific journals, and predict their potential usefulness. This problem is particularly prevalent in medical studies. A study to determine the cause of cancer is science; development of cures is technology. The U.S. has excelled, traditionally, in technology and engineering; Europeans have led the development of science.

Technology today seems to have shifted focus toward control of natural processes to achieve future benefits. Technology is a complex process affected by uncertain social, political, economic, and destructive natural phenomena. So prediction and control of future events are uncertain. The history of technology forecasting has been quite dismal. From our perspective today, it seems impossible that no one at the Chicago Exposition in 1893 foresaw that cars and airplanes would change the structure of society in the 20th century. Is it any more likely that those today forecasting technology fifty years ahead will be any more successful?

Because science is difficult to understand, those who do are held in high esteem by the public For the most part, scientists are specialists in narrow fields of studies without special knowledge of how technology will be utilized in the future. Scientists are not, necessarily, bestowed with superior wisdom and knowledge. Scientists can be wrong.

Scientific knowledge developed in the 20th century has come into conflict with moral and religious beliefs. Examples include the creation of nuclear weapons capable of destruction of all living things, and the creation or modification of living things by gene manipulation. The development of artificial intelligence, virtual reality, and drone aircraft could lead to dangerous consequences for society. People might wish to back off from some of these discoveries, but knowledge once acquired, can be difficult to contain. Knowledge must be managed to optimize benefits and minimize unwanted side effects.

Science & Technology

Although science continues to explore the universe, more attention is now directed to understanding life itself and the world we inhabit. In the twentieth century, science learned a great deal about the universe, the design of materials, and the functioning of nature. Science still has much to learn, however, about human beings, plant growth, and planet Earth – oceanography, climatology, and geology. The study of Earth is still in its infancy.

As science expands the bounds of knowledge the cost of experimentation rises. The costs of exploration of the solar system are in the billions of dollars, and building and operation of particle accelerators exploring the composition of the atom have similar costs. Can the U.S. continue to afford such expensive scientific programs as expenditures are rising for social programs and maintaining the nation's infrastructure?

Part V

The 20th Century Legacy

Humanity's Struggle with Human Nature and

The Evils of Technology

13. The 20th Century Legacy

Science and Technology

The 20th century was the most remarkable period in the history of human civilization. Packed within little more than one hundred years was the mapping of the universe from the interior of atoms to the ends of space. Science blossomed. The nature of matter was defined. Power was created from the energy of chemical combustion and from the splitting of atoms. Power now lights the world and sends vehicles around the Earth and into outer space. Vehicles of all sizes transport people and goods throughout the world. Information and images fly around the globe at the speed of light.

The environment was exploited as never before to create new metals, ceramics, and composites that made living more comfortable and stimulating. The Carbon Age was born as synthetic materials were created that supplemented natural materials for construction and clothing. Carbon products are now critical components of nearly all materials used today. New medicines and biocides were invented to fight disease and improve safety of food. Anesthetics and analgesics relieved much of the pain of living.

Technology has imposed great dangers on the world in the guise of progress; unintended consequences. Most people are aware of the horrendous dangers posed by nuclear weapons, and attacks on civilians by terrorists armed with plastic explosives, poisons, and automatic rifles. The cyber world is plagued by hackers and misuse of social networks. Fewer, perhaps, realize the dangers posed by world travel and commerce that spread contamination plagues, and undesirable species wherever people and goods go.

The greatest danger of all, however, that is not obvious is the impairment of communication and understanding even as messages transit the world at the speed of light. Listeners are bombarded with data, pictures, and opinions, without being able to understand neither the message nor the truth of the information. Not even pictures can be assumed to be true. We are all specialists in a world of specialization, unable to understand other specialists with different experiences. So we readily accept the word of others whom we believe to be more knowledgeable. In the less complicated world of the past, people could understand most things and defer to gods for guidance of the unknown. Now people cite the word of "scientists" to guide them, but scientists too are specialists and lack divine judgement. Understanding the workings of the specialized society that technology has created is one of the greatest challenges facing humanity.

Society and Government

The population of the world increased by many billions of souls, thereby increasing pressure on the environment and competing for space with wild animals. Populations, formerly spread thinly across rural landscapes, expanded with many millions living in large megacities. The structure of societies around the world changed as cultures adopted technologies and were brought together into intimate contact. Nations were formed by combining neighboring cultures and by dissolving empires.

Growing demands for power and materials for expanding populations place great stress on the environment. Pollution abatement and conservation of resources have become critical issues for mankind.

Society changed markedly in the U.S. as the influence of Christian values and property rights waned and concern for individual rights and environmental protection grew. Society seeking immediate satisfaction borrowed heavily from the future. International commerce and trade grew as the benefits of technology spread around the globe. Enormous financial institutions evolved to finance the expanding world commerce. Governments grew to exert greater effects on the lives of citizens.

Education changed greatly in the 20th Century, both in methods of instruction and increased government influences. The natural way to learn is through experience - trial and error and mimicking others. In earlier times pupils learned the basic skills of reading, writing, and arithmetic by practice, and they learned working skills by apprenticeship and on the job training. Then schools proliferated as society changed to require higher levels of education and teaching became a major career. Focus changed from students learning what they wanted or needed to teachers telling them what they should know. Freedoms are threatened when governments tell teachers what to teach the youth.

Generations have grown up in a cyber world of imagination, seemingly, unaware of the material world of facts and truth around them. People expect technology will create a better world in the future. Technology, however, cannot provide a magical solution to the nation's present social and cultural problems. The cultural bonds that united peoples for centuries are under stress from conflicting ideas and opinions. Society has become highly polarized and hard pressed to manage the innate destructive behavior of its citizens.

The nature of organizations is to grow, and bureaucracy grows by expanding regulations. While some laws and regulations are necessary to prevent chaotic markets, too much regulation impedes commerce. Regulations now exceed the laws passed by Congress in numbers and costs to society. Curtailing onerous regulations will foster economic growth by the private sector and justify reducing the size of government. The huge size of government has overwhelmed its economic base. The federal government and many states and municipalities are operating with large and growing deficits.

Inefficiencies are built into our republican form of government when federal, state, and local governments address the same issues. Initially, the federal government was intended to deal only with foreign affairs, with protection of its citizens abroad and at home, and with commerce which affected everyone. Beginning with the Great Depression in the 1930s, the federal government has assumed a major role in social issues as well, sometimes forcing programs on reluctant states. A uniform policy is not always necessary for the component states of a republic. What benefits people in the urban Northeast, for example, may be inappropriate for people in the rural Central-States.

Government and education costs have grown unduly in recent years. High salaries, and benefits negotiated by government unions is one factor. Politicians have been loath to deny union demands. Increased employee benefits in government put added costs on taxpayers and reduce economic competitiveness. Unions created in the 20th century to combat managements' dictatorial control of labor in the manufacturing and mining industries, are inappropriate for government employees who provide critical services to the public. Government cannot allow essential services to stop. Past federal governments in emergencies have refused to accept strikes by unions and forced continuance of essential services by replacing strikers with military personnel. Too often, however, politicians find it easier to accede to employee demands than to contest them.

In the 20th century the United States became the most litigious nation in the world. Accidents, mistakes, and ignorance have been "criminalized" in that someone must be held accountable and pay compensatory costs for every incident. Legal expenses and liability insurance to protect companies and individuals has become a major economic cost to society.

Government and the Economy

Prior to the 20th century, manual labor was the basic unit of society and the economy. Manpower or critical resources were the limiting factors in production. Technology and automation created the need for specialists and caused a surplus of unskilled labor. A major issue for society has become keeping working populations gainfully employed and providing relief for the unqualified. Economic theory has not kept up with this change in the

The 20th Century Legacy

functioning of society from reliance on brawn to dependence on brains.

Commerce is now international; people and goods move freely throughout the world; the U.S. no longer enjoys economic or technical advantages over the rest of the world. To prosper, the U.S. must bring its costs in line with those of other countries. Withdrawing behind our borders and going it alone is not an enduring option.

In the past value was created by human effort; items were valued by the amount of work needed to create them. When children wanted something, they were expected to earn at least some the money to pay for it. They then cherished and protected it. Machines now make items so cheaply they have little value for the owner. Goods are trashed rather than repaired. Few save any more to buy a car or a house; they lease or borrow to possess them, but do not own them. Hence, there is little pride of possession. We have become in effect renters of goods, not owners. The culture today has deviated greatly from that of the nation's forefathers, who cherished freedom and valued ownership. Debt was frowned on because it reduces individual freedom.

Misuse of Technology

Humans are very inventive and quick to offer opinions and solutions to problems. However, people's knowledge of issues is limited by their experiences. The functioning of science and technology is a mystery to most of the populous; so it is easy to exaggerate the knowledge and wisdom of scientists we can't understand. Often those with the visions are confident science or technology (most don't understand the difference) will make their visions come true. NASA began as a program to find peaceful uses for technologies developed from the military space program, but expanded into cooperation among nations for scientific exploration of the universe. With each successive project, costs increase and benefits for society are fewer. Costly projects to build high speed trains have been undertaken because other nations have them and the U.S. doesn't, not because they would improve U.S. transportation. There is little consideration of costs and value for the nation and its citizens in such projects.

Scientific discoveries and technical achievements can be impressive, yet provide no economic benefit for society. Novelty alone is not enough for commercial success, no matter how spectacular the development might be. Sending humans to walk on the moon was a spectacular achievement, but is little practical use to people on Earth. It is hard to place economic value on scientific knowledge.

The crusade by environmentalists over the past forty years to replace fossil fuels by electricity generated from renewable sources has been very costly to society and has misdirected pollution abatement efforts. The government mandate to use alcohol in automobile fuel has upset the corn and agricultural markets, and contributed to higher energy costs. Government subsidies and regulations to force electricity generation from wind and from solar radiance also have had a disruptive effect on the nation. Electricity is very useful, but it is not suited to all power needs and it is often more costly than alternatives. Misuse of technology by government, likely, is a contributor to the growing government deficits of recent decades and is a costly burden on the economy.

Proper use of technology requires the wisdom to distinguish between scientific truths and scientific theories or opinions; and between science which seeks knowledge and technology that provides useful economic gain. To make the 20th century legacy a happy one, humanity should return from visions of space and virtual reality and unite to address the current economic and social issues of displaced populations, under employment of young people, threat of economic collapse, and cultural conflicts. Wants and desires must be brought into balance with the earning power of the nation.

Humanity's Struggle with Human Nature

When the 20th century began, the United States was proud to be called a melting pot as immigrants from many nations were assimilated into an English-speaking, Christian nation. Diversity was accepted and respected. People could joke about their differing ethnicities as they worked together becoming Americans. Catholics, the various protestant sects, and Jews, warily coexisted. Segregation and prejudice toward black people persisted through much of the nation impeding their economic and social assimilation. Vestiges of this centuries old issue remain for some people today. Freedom, independence, hard work, and justice for all as expressed in the "Pledge of Allegiance", were the values and aspirations of the U.S. culture. "God's Law" described in the Jewish and Christion teachings are the foundation of the U.S. government and defined the American culture.

In the decades following World War II, society changed to celebrate diversity, instead of the diverse factions adapting to fit into an American society as they did in the past. Disrespect for those with opposing views has grown. Drug and sexual abuse have become widespread. Gangs flourish in the inner cities, competing with each other for territory and control of drugs.

The diversified society today has become very sensitive to derogative words and action by others. Need for "Proper Speak" accentuates diversity. The religious foundation of the country is under attack as atheists attempt to erase "under God" from the American heritage. Competing special interest groups, with little respect for their opponents, now resist working cooperatively to resolve issues. People often judge past social behavior by modern norms, and expect others to atone for bad actions of their ancestors. Terrorism by disgruntled or deranged citizens is a growing problem. It is difficult to create unity with people focused on their differences.

The United States has become the "Divided States". It is divided between those living in the populous urban states along the east and west coasts and those living in the numerous states in the rural midsection. It is divided between those in favor of government control of the economy and those favoring property rights and the freedom of private enterprise to create wealth. It is divided between those valuing equal opportunity and meritology in organizational advancement and those favoring representative placement of minorities in organizational hierarchy to mirror the ethnic populations. It is divided between the religious and the secular.

What happened to change the U.S. culture?
The U.S. enjoyed increasing economic prosperity following World War II as benefits of science and technology were adopted by society. It was also a time of fear of nuclear destruction as the Cold War with the U.S.S.R., the battle between Capitalism and Communism, was waged. Some people built bomb shelters in their yards; even the federal government constructed a complex shelter deep under the Greenbrier resort in the mountains of West

Virginia to house the government in a crisis. Young men were called into the military to fight in distant lands to curtail the spread of communism and promote democratic ideals. U.S. culture and commercial influence spread through the free world.

Then an anti-war, counter culture movement developed that renounced materialism and U.S. cultural values and focused on the human condition and introspection. Advocates (notably, beatniks, and hippies) sought to explore body and mind and free their personal self from moral and legal restrictions. Soul searching was enhanced by loud music (first jazz and then rock and roll) and psychedelic drugs. A Free Sex movement evolved that made all kinds of sexual behavior publicly permissible, overturning long held moral restraints. Gay rights including, gay marriage became law. Ignoring biological and physical differences, women were freed to compete with men in formerly male dominated occupations, too often leading to unwanted sexual advances. National borders and personal property rights were no longer respected. Even illegal immigrants now claim a right to live in the U.S. and enjoy freely the benefits of U.S. citizenship.

Weakening of the importance of family as the basis of society began with shift from rural to urban living. Then sexual behavior changed from an innate drive to procreate to a casual sport. Development of birth control pills, spread of abortion practices, and glamorization of sexual behavior by media and movies changed social behavior. The women's liberation movement lured women away from family responsibilities. Now it seems children mainly serve to satisfy maternal and paternal instincts, and too often, family values no longer are the main influence for maturing youths.

In this anti-materialistic society, little attention is given to commercial enterprise and economic development. Wealthy people and industrial corporations have become the tax targets of government. The new society expects immediate gratification and enjoyment with little concern for the future.

Nationalists vs Globalists

The U.S. has become divided between those who retain the cultural values of the Christian founders of the nation and those who seek change. The former now are dubbed nationalists and accused of having excessive patriotic zeal, bigotry and prejudice against those who disagree with them. The new anti-materialistic opposition, called globalists, disavow national boundaries and personal property rights. They favor nations opening their borders to immigrants seeking a better life. The split is between those who believe a culture composed of people united by common values should be the basis of society, and those who think society can be based on individuals. This is celebrated as Democracy. In effect, globalists deny the importance of a culture base for nations.

Culture Matters

The young 21st century was bequeathed a society facing broad social issues, which governments are ill-equipped to resolve. With the lessening of the influence of Christian - Judeo family values in the 20th century, governments have taken on care of those unable to care for themselves, usually with financial aid. Financial aid alone, however, does not solve human behavioral issues such as, idleness, homelessness, violence, sexual and substance abuse, and loss of common cultural values. It is natural for humans to live in families until maturity where they learn values and independence. Formerly, families took care of those family members who couldn't care for themselves. It was common for three generations to reside together, but it is rare today for grandparents to reside with their children.

Humans have an inherent need to belong to a group with common interests (family, race, region, club, school, religion). Language, genetics, race, and religion are basic determinants of culture. Families and tribes, traditionally, establish the values and enforce behavioral rules that characterize their culture. Prior to 1800 humans around the world lived in familial communities and principalities. Then in the 19th century nations were formed in Europe by combining adjacent principalities (notably in Germany and Italy). After defeat in World War I, the Ottoman Empire was arbitrarily split into nations without concern for the tribal boundaries. The relocation of expatriate

The 20th Century Legacy

Jews to Israel in Palestine after World War II, stirred continuous cultural conflict with neighboring Muslims. Many of the Wars of the 20th century in Asia and Africa were rooted in cultural conflicts.

History teaches that a nation combining conflicting cultures is unstable and will not persist. Yugoslavia and the U.S.S.R., which both fragmented into their component cultural entities, are 20th century examples. The European Union is in danger also of breaking into cultural units.

The social issues now facing the nation are mainly cultural, not political. Cultures provide the social controls that curtail destructive human behavior, not government mandates. Prohibition laws failed to prevent liquor abuse; present laws are not curtailing drug abuse nor smoking tobacco by minors; and laws are unlikely to curtail weapons abuse and bring peace to inner city neighborhoods. Humanity is losing the struggle with human nature. Lack of common values is the root cause of the present social problems that divide the country.

The 21st century inherited a world of imagination, of cartoons, of exciting visual and audio effects, of space wars and extraterrestrials. Most waking time of people is now spent watching computer and movie screens, texting, listening to music, and exploring "Virtual Reality." Exercise is now workouts in gyms and spas to maintain the health of over-fed and under-worked bodies.

It is widely believed today that democracy is the best form of government, but this was not the view of the Nation's founders. James Madison, the chief writer of the Constitution, viewed democratic assumptions as seductive illusions. He considered "the people to be an ever shifting gathering of factions committed to provincial perspectives and vulnerable to demigods with partisan agendas." [8] The Constitution was designed to prevent the dangers of mob rule.

The Constitution was carefully crafted to create a central government that would insure the freedom of the people and the independence of the states with various cultures, while protecting against the dangers of a dictatorship. Those protections were removed in the 20th century, as the executive branch has assumed dominance over the Congress and the state governments, and only the Electoral College remains to stand in the way of an unruly democracy.

The presidential election of 2016 and its aftermath featured a loud but unsuccessful movement to bypass the Electoral College, that last protection of cultural freedoms. This election, probably, was the first time in American history that the losing side continued public protests and demonstrations against the policies of the winning party after the election was decided, and refused to participate in legislative functions thereafter. For a nation that stresses the peaceful transfer of power in democratic elections, the election of 2016 was a poor example.

Technology Effect on U.S. Society

The 20th century was a time of great economic and social changes resulting from thousands of technical innovations. To understand the impact of these innovations, it is instructive to consider the changes in five general areas: wire grids, wireless communications, horseless carriages, flying machines, and electronics.

Wire Grids
The century began with electric wires being strung to bring light to cities and buildings. Soon high voltage power lines traversed the nation and most buildings and homes were fitted with electric wires. New industries were created to manufacture insulated copper wires, transformers, motors, and switches and to manage the electrical grid. Companies were founded to create electrical machinery and electrical appliances, expanding the markets for electric power. Electric power is essential for modern society.

While utilities were bringing electric power to the nation, Alexander Bell's company AT&T was stringing telephone lines to connect homes and businesses across the nation. Telephone cable was laid on the ocean floor to connect the U.S. with European countries. A large industry developed as telephones became indispensable for homes and businesses.

The 20th Century Legacy

Wireless Communications As telephone lines were spreading across the nation, wireless radio was developing to reach those that telephone wires could not reach. Radio communication was critical for emergency and military personal, ships, and later, airplanes. For many years, radio broadcasts provided audio news and entertainment for the nation. With greater transmission power and improved reception equipment, the range and quality of broadcasts expanded. Television was introduced in the 1950s, first in black and white and later color. The subsequent introduction of satellites and microwave booster towers, digital cameras and cell phones, extended the reach of communications worldwide. Microwaves have expanded the reach and convenience of communications, but weather and solar activities can adversely affect transmissions. Optical glass fibers often replace electric wires for the rapid transfer of pictures and information.

Horseless Carriage

The development with the greatest effect on society in the 20th century was invention of automobiles that replaced horses and carts for local travel. Unlike horses, however, autos required smooth roadways; so the economic story is the construction of roads and the modification of towns to accommodate and service autos and passengers. Cars allowed people to live miles from their employment leading to suburban housing and an enhanced sense of personal freedom. Until the interstate highway system was built in the 1960s, auto travel was mostly local. Motor homes fostered travel throughout the nation. The many businesses and industries that evolved to support motor vehicles are major factors in the U.S. economy.

Flying Machines

Airplanes were the last to have a major economic effect on society and the way we live. Airplanes were invented concurrent with automobiles, but aircraft had little commercial effect until the last half of the 20th century. They were developed first for military uses by European countries in early 1900s; the U.S. had little interest in military aircraft until World War II. Planes were developed in the U.S. first for barnstorming and racing exhibitions and for setting flying records. The U.S. became the major producer of war planes during the World War II and of commercial aircraft in the post war years

The first commercial passenger aircraft had limited capacity and were relatively slow. The industry grew rapidly in the years following World War II as planes were developed with greater passenger capacity and municipal airports were built. Scheduled flights connected the larger cities, and nations competed for international routes. As commercial aviation expanded smaller private and business aircraft were developed that competed for air space and airport accommodations. Government stepped in to establish rigid flight rules and to monitor and control the airways. Proliferation of low flying helicopters and unmanned drones further complicate use of the air space. Air planes have become the preferred method of travel for distances over a few hundred miles at the expense of railroads and ships. Operators of airports struggle to keep up with expanding service demands of new aircraft models and the growing volumes of passengers.

Electronics

The latest of the 20th century innovations were electronics, which have affected the way people behave but have not changed the physical structure of their lives as did the previous technologies. Electronics marginalized the value of labor and enhanced the need for specialized knowledge. Service, hospitality, and entertainment industries have become the major economic factors, replacing agriculture, mining, and manufacturing. Automation, computers, data processing, and graphic effects have enhanced our leisure time, but have done little to grow the economy.

Concluding Remarks

The 20th century created great advances in scientific knowledge and spectacular scientific tools. It also brought a burgeoning world population and widespread desecration of the environment. Spectacular scientific achievements have led people to exaggerate the capabilities of technology and science and to attempt to design the future to their liking. Science, however, has not made people smart enough to create the future they might like. People should recognize that technology must not only be useful but it must also be cost effective to provide a better life. Solar energy, while useful, is relatively costly and not suitable for many energy applications, nor is solar likely to be the key to controlling climate change.

Instead of funding science for further costly exploration of the cosmos, the nation would be better served by encouraging colleges and universities to emphasize study of the Earth and biological sciences where far less is known. Better understanding of the composition of Earth and the forces that affect it are needed to protect and utilize the environment of our only planet for the benefit of all.

A better future comes from resolving current economic and social issues and deriving more value from the environment.

References

1 - Al Gore, "Earth in the Balance", Houghton Mifflin, 1992

2 - Thomas Donlan, Barron's Editorial, January 9, 2012

3- Al Gore, "The Future: Six Drivers of Global Change", Random House, 2013

4 - Robert J. Gordon, CEPR Policy Insight #63, 2012

5 - Tyler Cowen, "The Great Stagnation", Dutton, Penguin Group, New York, 2011

6 - Angus Deaton, "The Great Escape", Princeton University Press. 2013

7 - Charles Esdaile, "Napoleon's Wars", Penguin Books(USA), 2009

8 - Joseph J. Ellis, "The Quartet", Alfred A. Knopf, New York, 2015

9 - Robert Piccioni, "Atoms, Einstein, Universe", Real Science Publishing

10 - John and Patricia Adams, "A Force for Nature", Chronicle Books LLC, San Francisco, 2010

11 - Al Gore, "Our Choice - A Plan to Solve Climate Change Crisis", Rodale Inc. Emmaus, PA, 2009

12 - Jon Gertner, "The Idea Factory", the Penguin Press, New York, 2012

13 - W. Cleon Skousen, "The 5,000 Year Leap", National Center for Constitutional Studies, Twentieth Printing, 2013.

14 - Thomas Donlan, Barron's Editorial, July 25, 2016

15 - Niall Ferguson, "The War of the World", Penguin Books, 2006

Acknowledgments

I want to thank the many talented residents of University Village Thousand Oaks who have welcomed my attempts to explain the workings of science and chemistry. Their questions and challenges are the impetus for this book. I want to thank particularly Bill Bang, Blaine Shull, Bill Schwartz, Eugene Motte, and Bengt Sohlen for their counsel and suggestions, Special thanks to Chuck Kircher for his guidance and assistance in publication and John Forster for editing.

I would be remiss if I didn't acknowledge the works of John Adams and the Natural Resources Defense Council, and the books of Al Gore for their exposure of the wide spread desecration and pollution of the land and their and attempts to correct the practices. Their writings were an inspiration for this book. In my view their crusade to alleviate pollution was derailed when, lacking understanding of the complexities of energy technology, they concluded that the only way to reduce pollution was to eliminate use of fossil fuels.

Finally, I want to thank my wife Mary Lou for her patience as the book slowly took form.

About the Author

In the 1950s it was said that 90% of the scientists who ever lived were then at work. It was the heyday of the development of the chemical industry. Frank J. Welch began his career as a research chemist in Union Carbide Chemical Company, one of that 90% of scientists. He contributed to the expanding knowledge of the chemical composition and structure of polymers acquiring many patents. His career shifted from the "Ivory Tower" of scientific research to management of technical programs for development of useful new products. Subsequently, he joined Avery Label Company as Director of R&D where he led programs to develop ecofriendly coating and printing processes and new products.

Welch grew up in Fresno, California, during World War II. He received a B.S. degree in chemistry from Fresno State College and PhD in Organic Chemistry from Stanford University. He now enjoys living in an active retirement community in Thousand Oaks, California, with Mary Lou, his wife for more than sixty years.

The 20th Century Legacy

The 20th Century Legacy

www.ingramcontent.com/pod-product-compliance
Lightning Source LLC
Chambersburg PA
CBHW051803170526
45167CB00005B/1865